GEOGRAPHICAL RESEARCH AND WRITING

GEOGRAPHICAL RESEARCH AND WRITING

ROBERT W. DURRENBERGER

SAN FERNANDO VALLEY STATE COLLEGE

THOMAS Y. CROWELL COMPANY

NEW YORK / Established 1834

L.C. Card 77-136033
ISBN 0-690-32301-8
DESIGNED BY VINCENT TORRE
Manufactured in the United States of America

Preface

Each academic discipline has its own body of literature and its own peculiar way of looking at the set of problems it chooses to investigate. It is the task of practitioners in each discipline to prepare beginners so that they can continue the researches and investigations already begun. This book has been developed as a guide to assist students doing research and writing in the field of geography. Although directed specifically to undergraduates preparing their first research papers in geography courses, it will also be useful to students doing advanced work in the field. I make no claim, however, that this is a comprehensive work on geographical research techniques. In particular, the book lacks any considerable discussion of either field or quantitative methods. Rather, I have attempted to present some general notions and information that will give students sufficient background to complete their undergraduate writing assignments satisfactorily.

To that end, I have sought, in Part I, to convey some idea of the nature of geography, of the kinds of problems that preoccupy geographers, and of the preparation that is required for one to become a competent researcher. In Chapter 2, I have discussed the ways in which research problems can be isolated and limited and research plans can be developed. Chapters 3 and 4 treat research techniques and manuscript preparation in some detail; however, students are best advised to ask their instructor to specify the style manual that he prefers or that the college requires for the preparation of research reports or theses.

Part II supplements and updates John K. Wright and Elizabeth Platt's *Aids to Geographical Research*, 2nd ed. (New York: American Geographical Society, 1947) and should be used in conjunction with it, as well as with the Association of American Geographers' *A Geographical Bibliography for American College Libraries*, compiled and edited by Edward T. Price, Jr., and Harold A. Winters (Washington: 1970).

In preparing *Geographical Research and Writing*, I have been guided by the conviction that problem identification, problem analysis, and the preparation of research proposals and reports are skills that every educated person must possess. Today, with the growing importance of research and development activities and written communication, geography majors will profit from a program that develops skills in these areas, no matter what occupation they eventually enter.

The material contained herein has been gathered from a number of standard sources, most of which are listed in Part II. In particular, however, the sections on library material and library use have been extracted from *The Library Handbook*, San Fernando Valley State College, and are used with the Library's permission.

The book has been critically examined by my colleagues and students at San Fernando Valley State College; additional comments, suggestions, and criticism for its improvement are most welcome.

R.W.D.

Contents

PART II

AIDS TO GEOGRAPHICAL RESEARCH

Contents

PART I

RESEARCH AND WRITING

CHAPTER 1

The Nature of Geographical Research

Research is big business today. It has been suggested that by the turn of the century one-half of the gross national product will be devoted to research and development. Although much of this money will go to applied research for the solution of problems of the community, the state, and the nation, considerable sums of money will also be allocated for basic research. Additionally, unsupported research will contribute greatly to our general fund of knowledge and to the solution of society's problems.

The line of demarcation between basic (pure) and applied (service) research is not a hard and fast one. Essentially, basic research involves the asking and answering of questions that do not involve immediate solutions of pragmatic problems. Applied research is generated not because of someone's intellectual curiosity but because some problem of society demands a solution. Much of applied research focuses on satisfying human needs, such as supplying better food, shelter, health, and societal institutions. Pure research seeks to fill gaps in our knowledge and to advance the frontiers of knowledge in the various academic disciplines.

Geographers participate in both basic and applied research. In

3

the field of urban and regional planning they are working on the problem of organizing modern society into manageable form. In the Quartermaster's Environmental Research Unit at Natick, Massachusetts, climatologists are studying problems associated with military operations under various environmental conditions. Geographers in retail merchandising organizations study the factors influencing the relative success or failure of stores in different locations, and geographers in private enterprises and in governmental agencies seek better ways to meet society's problems.

Geographers frequently discover significant problems through reading and from discussions with scientists in their own and other fields. Professional meetings, where individuals hear about the most recent findings of others in their own and adjacent fields of investigation, often stimulate new research or shed new light on old research. Professional journals publish research findings and serve as a forum where conclusions may be tested.

Early Methods of Acquiring Knowledge

The study of the earth and its inhabitants began long ago. Early man looked about and asked such questions as: What lies beyond the horizon? Where does this river rise? Where is its mouth? Is the world round? Is it flat? Why are some people dark-skinned and others light? Why do people live where they do and in the way that they do? Curiosity impelled man to explore the four corners of the globe seeking answers to these questions. And it seemed that each new discovery or new set of facts led to new questions which demanded new answers.

Thus, the arts of observation and description became highly developed among most primitive people, and the science of geography (from the Greek words meaning *earth description*) came into being. Observations of natural phenomena gave rise to attempts at explanation, and proverbs accounting for significant natural events

4

became incorporated into the lore passed from one generation to the next. But until men had accumulated a fund of knowledge about a given problem, the questions they asked either went unanswered or were explained as due to supernatural forces or to magic. Thus, the Greeks and Romans held the gods responsible for much that went on in nature—the rising of the sun, the wind, the rain, and the good harvest. Witches, sorcerers, and magicians accounted for actions of human beings that were out of their usual pattern of behavior.

Many legitimate concepts were formulated by the thinkers among primitive men and from generation to generation were passed on by the elders of the tribe as revealed truth. Thus, priests and teachers assumed an important role in society, and succeeding generations accepted the accumulated lore because it was presented by an authority. It is obvious that this method of acquiring knowledge and passing it on is still an important one.

However, the number of observations that one individual could make were few, and the means for disseminating knowledge were limited. As a consequence, many of the concepts developed by individual philosophers of the ancient world were incorrect. Thus, Thales, a Greek, thought that water was the universal substance of which everything was made. He probably arrived at this conclusion as a consequence of observing that all life forms required water. Other Greek philosophers concerned with the question of the nature of matter reached different conclusions based on their own experiences and observations. Anaximenes believed that air was created prior to water and so was the basic unit of all matter, and Heraclitus of Ephesus said that fire was the source of all things. Thus, many erroneous ideas were passed along with the truth. The idea of the flat earth persisted even after Magellan circumnavigated the globe. In time a number of individuals became dissatisfied with the results of the empirical-inductive method of reasoning utilized by these early philosophers and they attempted to reach conclusions in other ways.

One of the most effective techniques that they developed was the use of a reasoning process known as the syllogism. An example is given below.

> All metals are elements
> Lead is a metal
> Therefore, lead is an element

Thus, if the two preliminary statements (premises) can be shown to be true, it follows that the third statement (conclusion) must be true.

The syllogism involved a form of deductive reasoning in which a particular fact or conclusion is deduced as a consequence of observed relationships between two generally known facts or principles. Thus, early theoretical-deductive reasoning was based on the development of simple models of reality and assumptions that all of the real world fit into these theoretical models devised by philosophers.

Anyone who has been involved in or listened to a debate can comprehend some of the difficulties associated with too firm reliance on this method of reasoning alone to discover the true answer to a question. Debaters or politicians will use only those facts that support their thesis in order to convince listeners of the truth of their arguments.

Most of the scientists at work in the Middle Ages mistrusted the theoretical-deductive methods as a way to arrive at the truth. Such individuals as Galileo, Leonardo da Vinci, Copernicus, and Bacon stressed the need to make many individual observations of an event or an object before reaching a general conclusion about that set of events or objects. Bacon, in particular, warned against attempting to formulate any solution (hypothesis) until all the facts had been gathered and examined.

The impracticality, however, of Bacon's warning is self-evident. If you were attempting to prove that all swans were white, it would be impossible to examine all swans. You would have to reach a conclusion in some other way. Other methods of finding answers to questions and of reaching conclusions evolved as scientists pursued the quest for additional knowledge about the world in which they lived. One example of the application of scientific method in the solution of a problem is found in the following passage by Charles Darwin.

My first note-book (on evolution) was opened in July 1837. I worked on true Baconian principles, and without any theory collected facts on a wholesale scale, more especially with respect to domesticated productions, by printed enquiries, by conversation with skillful breeders and gardeners, and by extensive reading. When I see the list of books of all kinds which I read and abstracted, including whole series of Journals and Transactions, I am surprised at my industry. I soon perceived that selection was the keystone of man's success in making useful races of animals and plants. But how selection could be applied to organisms living in a state of nature remained for some time a mystery to me.

In October 1838, that is, fifteen months after I had begun my systematic enquiry, I happened to read for amusement "Malthus on Population," and being well prepared to appreciate the struggle for existence which everywhere goes on from long-continued observation of the habits of animals and plants, it at once struck me that under these circumstances favourable variations would tend to be preserved, and unfavourable ones to be destroyed. The result of this would be the formation of new species. Here then I had at last got a theory by which to work. . . .[1]

Darwin had used the information that he had gathered up to a certain point to develop a probable solution (hypothesis) to his problem. He then conducted further research to verify his hypothesis.

This is essentially the way in which most scientists of today operate. They identify the general problem to be investigated; gather facts and information pertaining to the problem; search the literature for related studies; delimit their problem; select one or more hypotheses; test and evaluate these hypotheses through the gathering of additional facts and information; select the final solution to their problem; and present evidence to support their case. Scientific inquiry, or scientific method, varies somewhat from

[1] Francis Darwin, ed., *The Life and Letters of Charles Darwin*, 2 vols. (New York: D. Appleton and Co., 1899), Vol. I, p. 68.

field to field but, in general, follows the above format and involves both inductive and deductive reasoning. Its application to the field of geography will be discussed in the next section of this book.

Modern Geographical Research

Geographical research is done by almost everyone, but the beginning of modern geographical research in the western world coincides with the voyages of discovery which followed the discovery of America by Columbus. Bit by bit the pieces of the world mosaic were put together by explorers and merchants who returned to their homelands with charts and written accounts of the places they had visited. At first this information was jealously guarded lest some other individual or state acquire dominion over a valuable land or resource. However, as greater numbers of missionaries, traders, and colonists ventured out of Western Europe, accounts of other parts of the world became more numerous. With the invention of printing, information about the people and places of faraway lands came within the grasp of people who were able to study maps and were eager to read the adventures of the travelers.

Thus, the field of geography owes much to that group of individuals who voyaged about the world, made maps and charts, kept logs, and returned to their homelands with firsthand observations of the places where they had been. In a sense, they represented the first modern geographers practicing a basic geographical technique —that of gathering data in the field. Their observations were the basic data used by the compilers of atlases and the writers of books —the armchair geographers of their day.

The knowledge explosion of this period sparked a flurry of scientific inquiry. As more became known of the distribution patterns of man and of climate, vegetation, landforms, and soil, questions arose involving cause and effect and the interdependence

8

of all living things. Also of importance to the mercantile nations were questions involving the natural resources of the world. Where were the new lands to which people might move and what material resources could be harvested or mined in them? Even deeper and more difficult questions involving the relations of man and nature and the ability of the earth to support more people were brought forward. In his essay on population published in 1798, Malthus looked forward pessimistically to the time when the number of people in the world would be too great for the limited resources available to them.[2]

It was in this kind of a world that geography as a modern academic field came into being. As might be expected, the two founders of modern geography were representative of the two groups of individuals that had done much to contribute to the fund of geographical knowledge up to that time. Alexander von Humboldt (1769–1859) was an accomplished explorer and scientist who gathered much of his own data in the field, made field maps and sketches, and returned home to write up his findings. Carl Ritter (1779–1859) was a typical armchair geographer who utilized the findings of others to prepare his compendiums of geography. Each made a significant contribution to knowledge, and each helped chart the course that geography has followed since that time.

As time went by more and more individuals in Europe and in this country identified themselves as professional geographers. Some went out to the field to gather their own data and make their own observations, many identified and classified the elements of the landscape as they became known to them. Most sought to explain the patterns that they found and to comprehend the processes operating to produce them. In doing this they sought to formulate general statements or laws explaining the patterns and relationships that they discovered. One such individual was Ar-

---

[2] One might note that some of these questions and problems are still with us, and the need for basic research for answers to these problems is still a pressing one.

nold Guyot, a Swiss geographer, who came to this country about the middle of the nineteenth century to teach at Princeton.[3] In his book, *Earth and Man*, first published in 1849, Guyot set forth some of the significant problems of the field of physical geography, which are as follows:

1. What *laws* govern the situation, extent, outlines, and relief of the *land masses?*

2. What is the *influence of the relief* of the continents upon the formation of their systems of rivers and lakes?

3. What is the cause, the extent, the connection, and the influence of the great *oceanic currents?*

4. What is the fundamental law of the *distribution of heat* upon the surface of the globe; what modifications of this law are observable; and how are those modifications produced?

5. What general *atmospheric movements* have been observed, and what is their cause, course, and influence?

6. What laws control the *periods, distribution, and amount of rain* upon different portions of the globe; and how is the existence of vast rainless regions in certain latitudes to be accounted for?

7. What general laws govern the *distribution of vegetable and animal life* upon the globe, and how are these laws related to the character and well-being of the human family?

By identifying problem areas and directing attention to them, Guyot helped define the limits of the new science of geography. By writing about the discoveries that they made on their own or read about in the journals of others, von Humboldt, Ritter, Guyot, and others helped diffuse knowledge about the earth and about the field of geography to all parts of the world. It was not long before departments of geography were organized in many countries, and professional associations of geographers came into being. With more geographers and with improved methods of disseminating knowledge new specializations and subfields of geography developed.

[3] Arnold Guyot, *Earth and Man: Lectures on Comparative Physical Geography in Its Relation to History of Mankind,* tr. C. C. Felton (New York: Scribner, 1873), p. 2.

As knowledge of the man-environment system increased and as observational and data collection techniques became more sophisticated, geographers developed more and better methods of analyzing the information acquired by scientists in all parts of the world; and as knowledge of the environment increased the questions that geographers asked became far more complex than those being asked in Guyot's time.

In the United States, many of the first professional geographers attempted to discover laws or rules governing the effects of the environment on man. Environmental determinism was the most active lobe on the frontiers of geographical research until the 1920's, when commercial and human geography began to occupy the interests of a greater number of geographers. In turn, regional geography moved to the forefront in the 1930's and 1940's as a consequence of Richard Hartshorne's monumental *The Nature of Geography* (1939) and the influence of Preston James, Chairman of the Department of Geography at Syracuse University.

Another major shift of emphasis in the field of geography in this country occurred as a consequence of the involvement of a group of geographers in governmental agencies during World War II. At the conclusion of the war a group of them held a conference at Hershey, Pennsylvania, to examine the nature of their contribution to the war effort and to make suggestions for improving graduate training for geographic research. Their report emphasized the need for emphasis on thorough preparation in the systematic aspects of geography. Many departments of geography did change their curricula and more geographers became interested in such subfields of geography as urban, cultural, historical, and theoretical geography.

The move toward greater emphasis on the systematic subfields of geography received additional support in a recent publication of the National Academy of Sciences–National Research Council, *The Science of Geography*. In it were presented the views of a select group of geographers on the research interests, methods, and opportunities of their discipline. The following quotation summarizes their idea of what constitutes the central problem of the field of geography.

Geography's Problem and Method

The Committee believes that geography, the study of spatial distributions and space relations on the earth's surface, contributes to treatment of one of the great problems of scholarship. This is a full understanding of the vast overriding system * on the earth's surface comprised by man and the natural environment. Indeed, it is one of the four or five great overriding problems ** commanding the attention of all science, like the problem of the particulate structure of energy and matter, that of the structure and content of the cosmos, or the problem of the origin and physiological unity of life forms.

The three great parameters for any scientific problem, albeit in varying dimensions and attributes, are space, time, and composition of matter. For the problem it treats, that of the man-environment system, geography is concerned primarily with space in time. It seeks to explain how the subsystems of the physical environment are organized on the earth's surface, and how man distributes himself over the earth in his space relation to physical features and to the other men. Space and space relations indeed impose one of the great mediators of the characteristics of any part of the system at any point on the earth's surface. As one of the major subjects concerned with spatial features on the earth's surface, and as the only one traditionally concerned with system interrelations within the space of the earth's surface, geography has a significant place in satisfying man's scientific curiosity.

Geographers have studied the space relations of the man-natural environment system for decades. At a time when few students in other branches of science were concerned with the relations betwen cultural phenomena and the natural environment, geographers were studying their spatial distributions and attempting correlation studies about them. Geographers' organizing concept, for which "spatial distributions and space relations" are a verbal shorthand, is a tri-scaler space. The scales comprise those of extent, density, and succession. The theoretical framework for investigating the man-environment system is developed from this basic concept. Central place (settlement) hierarchy, density thresholds, and diffusion theory are examples of theoretical constructs serving specific research.

Geographers believe that correlations of spatial distributions, considered both statically and dynamically, may be the most ready keys to understanding existing or developing life systems, social systems, or environmental changes. They further believe that geography has made a significant contribution in the past to the foundations of knowledge needed to understand subsystems of the man-environment system. Progress was gradual, however, because geographers were few, rigorous methods for analyzing multivariate problems and systems concepts were developed only recently, and few branches of science were committed to study of the man-environment system.

. . . . .

The situation now has changed enormously. Many aspects of systems theory and study techniques have invaded all of the traditional social sciences, biology, and geography. Rigorous analytical approaches offer a common ground for communication with one or more groups in every pertinent subject. Spurred by world events like the unprecedented increase in population, all science has recognized the overriding problem, the imminent need for understanding the world-wide man-environment system. But even as our capacities to analyze and eventually to understand the man-natural environment system have increased, the problem itself has increased vastly in its proportions. More than ever before there is a social urgency for effective research on it. As Hubbert has aptly shown, the present exponential increases of population and materials consumption in the world cannot continue for very long.

The members of this Committee believe geography has reached a critical stage of opportunity that derives from: (1) the now vital need to understand as fully as possible every aspect of the man-natural environment system, including spatial distributions, throughout the world; (2) the development of a common interest among several branches of science in the overriding problem and its spatial aspects; (3) the development of a more or less common language for communication for the first time among all the pertinent branches of science through mathematical statistics and systems analysis; (4) the development of far more powerful techniques than ever before for analyzing systems problems, including spatial distributions; and (5) a backlog of spatial experience which geographers have accumulated from their spatial perspective and their past dedication to study of the man-environment complex.[4]

* System—A functional entity composed of interacting, interdependent parts. A subsystem is a system which is a part of a higher level system.
** Overriding problem—A transcendent issue covering many of the problems customarily treated by scientists in theory, experiment, and observation.

[4] *The Science of Geography*, Publication 1277 (Washington, D.C.: Division of Earth Sciences, National Academy of Sciences—National Research Council, 1965), pp. 8–11.

In another part of this document members of the committee further define the nature of the field of geography by identifying some current research problems in the subfields which they are concerned with. Some geographers have objected to this report and have suggested that a field of study is best defined, not by what a group of experts in the field say it is, nor by an etymological analysis of the name, nor by its logical place among the sciences, nor by its place in the curriculum, but rather by the published and unpublished works of the members of that discipline.

Pattison, in examining the literature of the field of geography, has found four recurring themes (see Appendix I): (1) A Spatial Tradition, (2) An Area Studies Tradition, (3) A Man-Land Tradition, and (4) An Earth Science Tradition. Almost all of the scholarly activities of geographers fall into one or more of these traditions.

Geographers have always been concerned with problems of earth measurement and mapping, of geometry and movement, of the covariation of elements occupying space. They are also concerned with questions arising from variation of the earth's surface. Areal differentiation was a central question for geographers in the days of Strabo and still is today when urban geographers are at work attempting to comprehend the morphology of the city. Areal differentiation has always meant explanatory description. Man has sought to understand the processes operating within the man-environment system which account for the things that he sees in the world about him.

Geographers have long been asking questions about resource utilization, about the interrelationship of man and his environment, about the effects or limitations of the environment on man, and about man's role in modifying the face of the earth. What has been done? What is being done? And, what should be done? Problems of environmental perception and of seeking alternate solutions to the use of earth resources are recently formed lobes on this sector of the geographical research frontier.

The Earth Science Tradition is perhaps the oldest research frontier. Aristotle in his *Meteorologica* pulled together all that was known about the natural world and the relationships that existed between natural events and their causes. In the United States at the beginning of this century the first departments of geography grew out of geology; the first president of the Association of American Geographers was William Morris Davis, a geologist. Today, geographers are at work attempting to understand some of the same questions that plagued the Greeks in connection with the natural world around them. True, the questions asked are more sophisticated, and the techniques for the collection and analysis of data are much better today than in the days of the Greeks, but nevertheless the questions are still unanswered.

The field of geography today is comprised of many subfields in which clusters of research workers are studying different facets of a single problem area. Individuals in each cluster may be identified by the works they have published and by the courses they teach. They know the others in the subfield and the work that is being done so that there is a constant flow of ideas and information not only within a cluster but from one cluster to another. Some departments of geography represent almost all aspects of the field; others only a few. Many of the traditional research clusters are listed below.

1. Cultural Geography
2. Historical Geography
3. Population Geography
4. Settlement Geography
5. Urban Geography
6. Political Geography
7. Resource Management
8. Agricultural Geography
9. Manufacturing Geography
10. Marketing Geography
11. Transportation Geography
12. Recreation Geography
13. Biogeography
14. Medical Geography
15. Climatology
16. Geomorphology
17. Geography of Soils
18. Hydrogeography (Water Resources)
19. Field Techniques
20. Cartography
21. Aerial Photography
22. Statistical Methods
23. Regional Geography
24. Geography of the Oceans

This list is not all inclusive nor are the subfields of equal importance. As the nature of the discipline changes and as new research tools become available new subfields will be added in the future.

As an academic discipline grows and matures, changes occur not only in the problems with which the practitioners of that discipline are concerned, but also in the methods which they use to pursue research. In the developmental period of their discipline, geographers, as well as most other scientists, were concerned with gathering, describing, and classifying data and using empirical-inductive methods to reach solutions to their problems.

Thus, geography has always been a dynamic field of study, constantly shifting its methods of inquiry as new tools and new ways of looking at the earth become available. Today, data collection by complex electronic devices which may be carried by satellites and data analysis by high-speed computers have resulted in rapid changes in the ways in which geographers study the earth. Theoretical geographers using theoretical-deductive methods involving the creation of mathematical models to simulate various parts of the man-environment system now carry geographical research far beyond the levels of simple observation and inductive reasoning utilized by primitive man. And, although there are differing notions among geographers regarding the direction the field should take, there is general agreement on what the major problems are.

Preparation for Geographical Research

Many students enter the field of geography rather late in their academic careers; their only exposure to the field may have been in an introductory course in physical, cultural, or regional geography. Therefore, their knowledge of the field and its subdivisions will be limited. For such students about to begin research on geographical problems the best preparation is a broad liberal education and a general understanding of the nature of the field of geography. In addition, the student should acquire a thorough understanding of the known facts and accepted concepts of the research cluster in which he desires to work.

One way to acquire a general understanding of the nature of geography is to examine that portion of the college catalog listing courses in geography and to read some of the brochures published by departments of geography which describe their programs. Additionally, an examination of course syllabi and required textbooks will enable one to discover more about the topics discussed in each of the courses.

In addition to the above, preparation for research in the field of geography demands the development of the following qualities and skills.

1. *Creativity* Admittedly, an individual cannot be taught how to be creative, but he can observe those who are creative and be on the outlook for new and original approaches to the solutions of problems. In other words, one can develop his imagination and ingenuity; students need to be aware of the great personal satisfaction that comes from making significant contributions to society.

2. *Proficiency in the Use of Library Materials* Although undergraduate students may have used the library while doing research for essays and term papers, many are still beginners when it comes to utilizing library resources. A student should plan to spend some time at the start of each semester familiarizing himself with the general location of books in the library. This may sound unnecessary, but in rapidly growing libraries changes occur frequently. In particular, the student should carefully examine the reference section holdings. New aids to geographical research are published regularly. Knowing about them will save time later.

3. *Proficiency in Field Methods* A knowledge of experienced geographers' methods of research will help a student immensely in learning to do his own research. In many instances such knowledge will prevent repetition of the errors of previous scholars. Earlier it was suggested that new technology has given geographers the equipment to evolve new methods of observing the world. However, field observation is still needed today, for example, to check observations obtained by satellite photography, and one still

has to interview individuals to get at certain types of information. Some field techniques developed in the past are still applicable today.

4. *Proficiency in the Use of Maps and Photographs*   Virtually all research papers in geography require illustrative materials to convey certain concepts or to amplify the written word. For this reason, geography departments ordinarily require their students to take a course in map and photo interpretation. One of the satellites circling the earth, *Eros* (Earth Resources Orbiting Satellite), is of great significance to geographers. Satellite photography yields useful information on such diverse topics as urban land use, the spread of forest and crop diseases, the location of fish-bearing ocean waters, the thermal pollution of river water, and a myriad number of other aspects of the man-environment system.

5. *Proficiency in the Analysis of Statistical Data*   At one time only a limited number of geographers felt the need for competency in quantitative methods. Today, scholars in most of the research clusters in the field make use of statistical analysis as one of their principal research tools.

The general availability of automatic calculators and computers makes possible the analysis of far greater numbers of bits of information than heretofore possible. The development and testing of theoretical-deductive models is now a part of virtually every discipline. Workers in only a few research clusters in geography depend entirely on library or field research methods.

6. *Mastery of One or More Foreign Languages*   Geographers pride themselves on presenting "a world view." Their discipline is international and most of the problems they are seeking to solve are universal. Yet, many students of geography in the United States do not realize the necessity of being able to read foreign languages. The absurdity of one's becoming an expert on the geography of Latin America without being fluent in Spanish should be obvious. Thus, every undergraduate geography major is encouraged to develop fluency in at least one foreign language.

Again, although many of the significant works published in forcign countries are translated into English, many more are not. If the advanced student or scholar is to keep abreast of new developments in his particular research area, he must know what his counterparts in foreign areas are doing. Languages not only must be learned in class but must be used continuously if they are to be useful research tools.

7. *Mastery of a Correlative Discipline or Disciplines* As an integrative discipline geography makes use of findings of scholars in other fields, and geographers often find themselves at work in problem areas with research workers in other disciplines. Thus, geomorphology involves geology; climatology involves meteorology, biology, and geology; urban geography involves sociology, economics, and political science; historical geography involves history; biogeography involves biology; hydrogeography involves engineering and economics. And there are still other areas also requiring contact with other scientists.

If a student is to work on a problem involving research findings from another discipline, he must understand the nature of that discipline and be familiar with its literature and research methods. Admittedly, such knowledge at the undergraduate level will be somewhat superficial, but a research worker should add to his knowledge of the correlative disciplines through attendance at lectures and professional meetings, and examination of the literature of the field as a matter of professional growth. In this way he will become aware of new advances in other fields of study and may find ways to adapt them to the field of geography.

Although all of the above listed qualities and skills are important facets in preparation for research in the field of geography, it may not be possible for some students to attain a high degree of competence in all as undergraduates. However, if one is planning to become a professional geographer he should develop as many of the qualities and skills as will be required to complete projects in the particular research cluster that he chooses to pursue. By keeping these goals in mind students will continue to build their competency to pursue research tasks outside of formal classroom activities. The attitudes and abilities thus acquired will serve them in good stead no matter what their life work eventually entails.

SUMMARY

Scientists conduct research essentially in one way. They identify the problem to be investigated; gather data pertaining to the problem; select one or more hypotheses; test and evaluate these hypotheses; and present evidence to support the solution. Geographers, in studying the facts of spatial distributions and space relations on the earth's surface, are engaged in data-gathering and classification and in developing theories which will contribute to their understanding of the man-environment system. Geographical problems fall into four recurring themes or traditions: spatial, area studies, man-land, earth science.

A student preparing for a career in geography should acquire a broad liberal education and an understanding of the field of geography. He should develop his creative capabilities and achieve proficiency in field and library research methods, one or more foreign languages, statistical methods, and interpretation of maps and photographs. He should also master one or more correlative disciplines to aid him in his search for interesting problems and to provide him with additional techniques to conduct his research.

Problems for Study

1. Write a definition of geography.
2. Discuss changes that have occurred in methods of acquiring facts for use in geographical research.

3. Discuss geography as systems analysis.

4. Identify the research interests of instructors in your department. How many research clusters do they represent?

5. Discuss Pattison's "Four Traditions." Which of these traditions do you feel are most viable at present? Defend your answer.

6. Examine the list of qualities and skills suggested as the best preparation for geographical research. Do you agree with the items listed? Are there other necessary qualities or skills? How well prepared are you to do research in geography?

CHAPTER 2

Identifying a Problem and Developing a Research Plan

From the time a student enters grade school until he leaves college he is required to prepare short essays and term papers. Most candidates for the master's degree must write seminar reports and a thesis, and most candidates for the doctorate must complete a dissertation. After leaving college one frequently has to write essays and reports in conjunction with work or for social and civic organizations. Skill in preparing a written report, therefore, is an invaluable aid throughout one's academic career and after.

Preparation of a research paper involves five different kinds of activities each of which requires a considerable amount of time and thought. The activities are:

1. Problem identification and delimitation
2. Development of a research plan
3. Data-gathering
4. Data analysis and interpretation
5. Report-writing

Each of these will be discussed at some length in the following pages. The first two are topics included in this chapter.

Selecting a Research Problem

Students, in most cases, select a topic very quickly, spend a considerable amount of time in research, and then rush to complete the paper just in time to meet the final deadline. The results are often disastrous. Good research and writing demand that sufficient time be spent in each phase of producing a research paper. The hours spent in identifying and delimiting the problem and in developing a research plan are as important as those spent in library and field research and in writing and rewriting the report.

Most college students are poorly prepared to identify research problems because, in learning facts in class, they are not made aware of the great gaps in the knowledge man has of the world in which he lives. However, new discoveries and new ways of looking at old problems suggest innumerable possibilities for further research. By examining the papers published in professional journals in the field of geography, a student will find a great variety of scholarly activities labeled as research. Some articles contain information gathered by the use of questionnaires. Others present broad generalizations based on observations gathered over years of research. Some represent reports on the status of research in progress; others are final reports of scholars' findings.

Types of Research

In many respects student research activities parallel those of older scholars. Both students and scholars acquire and interpret information in a number of ways and choose an acceptable form for presenting their findings. In the case of the mature scholar the nature of his finished product is determined in part by his own abilities and inclinations, but it is also structured by the editorial stipulations of the journal to which he submits his work.

Students enrolled in undergraduate courses in geography will ordinarily be asked to write research papers that fall into five general categories: (1) regional synthesis; (2) topical or systematic analysis; (3) evaluation of techniques, theories, or models; (4) literature review; (5) problem solution.

The first two categories involve research reports entailing description and interpretation. Students generally go into the field or to the library to gather information, which they then assemble according to some model established by their instructors or according to a format followed by a mature scholar in their field. In each case, the description of a region or of some systematic aspect of geography is followed by an interpretation or explanation of the assembled facts. An example of regional synthesis would be a paper on the Imperial Valley. A paper on some aspect of the petroleum industry would represent topical or systematic analysis.

The third category results from the continual efforts of researchers to develop better tools for the acquisition of data and better methods for the analysis and synthesis of the data that they accumulate. Reports discussing research tools and techniques may arise as by-products of previous research or they may be developed by researchers who devote their efforts to developing ways to utilize new data or devices.

Thus, the increasing use of computers and satellites has resulted in research on ways that geographers can make use of these new tools, and has made it possible for individuals interested in theoretical geography to develop and test mathematical models. The reports derived from such efforts are representative of the third type of research accomplished by geographers.

The fourth category, the literature review, is a paper that almost every professional geographer has written at one time or another. It may deal with a single book or monograph, or it may examine the literature on a specific region or topic. Special types of research reports involving a review of literature are discussed later in this chapter under the heading "Problem Analysis." Such reports focus on the identification of problems that have high potential for additional research.

Reports in the fifth category, problem solution, are ordinarily fairly straightforward accounts of the nature of the problem, suggested solutions to the problem, and advantages and disadvantages of suggested solutions. The form that such papers ordinarily take will be discussed in a later chapter.

Identifying Problems

In general, the best kinds of problems for a student to pursue are those arising from his own curiosity about something encountered in daily life or in a class. He is likely to do a better job of research on a topic that he is sincerely interested in than on one that is arbitrarily imposed. When a person becomes convinced of the need for solving a particular problem or when he feels an overpowering interest in finding the answer to a puzzling question, he not only derives great pleasure from his work, but generally is successful in completing his research and obtaining answers.

However, interest alone in a research problem is not sufficient. One also needs to have a clear understanding of the problem area involved; he needs to understand the present state of knowledge in that area and be aware of the work that already has been done on the problem. There are various ways in which this understanding can be acquired. One of them is discussed below.

PROBLEM ANALYSIS

In recent years, research analysts employed in government agencies and in research and development firms have developed a technique to help them define research problems. This technique—problem analysis—involves a review of the literature in the problem area and has three principal objectives:

1. To identify individual facets of the principal problem area and to discover related problems.

2. To ascertain the present state of knowledge about the problem.

3. To determine areas of needed research (unsolved problems).

This technique can be applied in many academic fields and is particularly useful to geographers. In addition to reviewing literature to learn the facts and the conclusions unearthed by others, students should learn to examine the nature of problems that are under investigation.

Problem analysis both inside and outside the academic world generally results in a report either summarizing previously completed research or evaluating current research. The author of the report must clearly define the problem area under investigation and indicate the scope and limitations of his study. His most important contribution is the identification of those aspects of the problem in which further research appears to be needed. Almost all government agencies and private research and development companies employ individuals whose sole task is to examine research reports and write contract proposals based on problem analysis techniques.

OTHER WAYS OF IDENTIFYING PROBLEMS

Quite obviously not all research problems are discovered by concerted analysis of literature. It is possible to develop your faculties so that you can recognize potential research problems when you see them or hear about them. Often a problem is revealed in conversation; wherever a difference of opinion exists, a problem exists. And the attempt to prove or disprove the truth of a statement or an idea involves research on the part of an individual seeking evidence to back up his own point of view. Often a difference of opinion between two mature scholars will spark a whole series of papers and published exchanges between them. Most professional journals publish criticism of a research paper and the author's rejoinder to such criticism at the same time.

Perhaps you work in a business or an institution that provides services to people. If you work in a hospital, you might be interested in finding out why the patients chose this hospital over another one. Is it because of the hospital's location? Because their doctor sent them there? What other factors are involved in the patients' decisions? Your findings might lead you to other questions. What are the principal factors that should be considered in locating hospitals? And, in actual practice, are hospitals located where you might expect them to be?

Students who work in retail stores have excellent opportunities to develop research papers associated with the problem of providing society with goods. Why are stores and shopping centers located where they are? Where do their customers live? What are the sources of the goods that are sold and how do they get to the store?

These are but a few of the questions that should come to students enroute to school or at work. Others may occur to them as they sit in a class or watch television at home. Most Americans are aware of the effects of the Asiatic monsoon on military operations in southeast Asia, but what about its effects on the people who live there? How do they adapt to the alternate periods of wet and dry conditions? Is culture a factor in the choices that they make?

You may discover other problems in your everyday experiences. If you commute long distances to work or to school, problems of transportation become meaningful to you. As a consequence, you might be interested in doing research on rapid transit systems to discover better methods of moving students to and from your college. Or you might come from a ghetto or travel through one on your way to school. Where are the limits of the ghetto district? How do you define them? If you lived in the ghetto would you define them in the same way you would if you lived outside the ghetto? What processes are at work to create ghettos?

If you are a keen observer, your journey to work or to school can become a form of geographical exploration. And, like the explorers of old, you may want to make charts and keep notes on what you see. As you travel, you will note variations in the landscape; perhaps repetitive patterns will appear. Such questions as the following may occur to you. Is there a rationale for the location of stores and gas stations? Are land use patterns around freeway interchanges the same everywhere? How do you account for the similarities or differences in land use patterns that you observe?

In class professors frequently allude to significant unsolved problems in their field. Issues and questions are raised but often go unanswered or are answered in an unsatisfactory manner. Each question is a potential research problem. Why not jot down ideas for research in a separate part of your notebook? Then, when you are asked to write a paper, you will have available a ready-made list of ideas to choose from and will not have to spend much time searching for a problem.

The unusual faculties of an individual born with exceptionally creative thought patterns are quite obviously not the only means of discovering worthwhile problems. Anyone through careful observations can learn to distinguish the important issues from the unimportant. A truly creative approach to a problem can be arrived at through well-planned and executed preparation as well as through a "flash of insight."

Questions to Ask about the Problem

Almost everyone starts looking for a research problem with a general idea of the nature of the topic he wishes to investigate. However, before attempting serious work on his topic, the student should do some additional reading and ask himself the following questions:

1. Am I truly interested in the problem?
2. Does it have significance?
3. Is it feasible?
4. Is the problem original?
5. What is the nature of the audience?

If your answers to the first four questions are affirmative, and you understand the nature of your audience, you are ready to move into the second phase in the preparation of the paper.

1. *Interest in the Problem* Evaluating your interest in a research project is a difficult task; in part, the success of the evaluation will be based upon your present knowledge about the topic. This is why it is generally desirable to select a research problem in an area in which you already have some knowledge. From personal experience and from reading you should be able to identify many research topics of interest to you. Unless an individual possesses some knowledge of his problem area he cannot have a real interest in his problem; unless he can develop a real interest in his problem it will be difficult to do a good job of research and writing.

2. *Significance* The topic you have chosen has significance if it extends your knowledge of a particular branch of geography or if it fills a gap in your general geographical knowledge. However, in preparing a research paper you must also consider your readers. Is the topic you have chosen likely to give them new knowledge or a new perspective on old knowledge? College professors appreciate a paper which presents them with new information or methodology or which presents a fresh view of an old subject. The best term papers are not necessarily written by those who have mastered the techniques of essay preparation but rather by those who have chosen to make a significant contribution to knowledge in a particular field.

3. *Feasibility* Student research efforts are generally limited by space, by data available, and by time. You may be asked to complete a ten-page paper within a period of two weeks. The nature of your research and thus your choice of topics are limited to what can be covered in the time allocated. Your research is also limited by the quality and quantity of the materials available. For instance, research on the agriculture of South Africa in a library without reference materials would be futile. And some topics should be excluded when there is no chance of doing field work.

Students should also learn to recognize the scale of the problems which they choose to investigate and the degree of precision involved in their solution. For example, a ten-page paper on a vast subject like the geography of the American West would not be long enough to contribute significantly to the subject, whereas a ten-page paper on the settlement patterns of the Humboldt River Valley might be quite important.

4. *Originality* Has someone else completed research on this topic? This question is important for two reasons. First of all, if there is already in existence a paper exactly like the one that you have in mind, there is no need for you to duplicate the research and writing already completed. On the other hand, if the other document is not precisely on your topic, it may furnish you with valuable information for your own paper or may serve as a model.

For the above reasons and to avoid being accused of plagiarism, it is necessary to complete a preliminary search of the literature prior to beginning preparation of your paper.

5. *Awareness of Audience* For whom are you writing this paper: for your fellow students, for your instructor, or for the general public? The type of audience will affect the approach to both research and writing.

In most cases, student papers are prepared in fulfillment of course requirements and are written for instructors in those courses. Therefore you should determine the instructor's desires and standards before starting work on the paper. He may want the paper prepared for presentation to other students or he may want it prepared for submission to a professional journal. Be sure that you meet his standards.

In doing the research and writing necessary for the completion of a research paper, you should keep in mind the reasons for this kind of activity. Term papers extend your knowledge of some aspect of the field of geography, increase your skill in the use of logical argument and literary expression, and are intellectual exercises which should enhance your competence in the use of library tools and field methods.

The Research Proposal

In the academic world the selection of a research topic generally involves a series of negotiations between the student and his instructor. In most cases these discussions proceed informally, although where the instructor has an unusually large number of students, he may try to structure the form in which students' research proposals are submitted. Whether formal or informal, the student should ordinarily proceed through the following steps before undertaking concentrated work.

STATEMENT OF THE PROBLEM

Even though he feels certain in his own mind of what he intends to accomplish by his research efforts, the student needs to commit his thoughts to paper. The precise formulation and expression of the problem is often more difficult and more important than the actual solution of the problem, which may only be a matter of applying an experimental or statistical method in a routine fashion. Understanding and stating problems clearly is for some the biggest hurdle in completing research.

In a research proposal the statement of the problem can ordinarily be accomplished in narrative form giving the reader sufficient background material to understand what the investigator is attempting to do.

STATEMENT OF HYPOTHESIS

By the time an individual is ready to state his problem he has accumulated a certain amount of information about the research area in which he is interested. In most cases he will have also developed some opinions regarding the solution of his problem.

These tentative solutions, or hypotheses, must account for all

31

the known facts in the case, and they must tie the facts together and explain their relationship to the problem. Ordinarily, several hypotheses must be formulated and tested before a solution is found.

In 1912 the climatologist Wegener suggested that much of the evidence for climatic change through the ages could be accounted for by the theory of continental drift. This idea was put forth in the form of a hypothesis. At first few American geologists gave much credence to his theory. In recent years, however, the overwhelming weight of evidence supports Wegener's hypothesis and the theory of continental drift may provide geologists with the answer to the problem of climatic change, as well as a number of other questions. Thus, a scientist at work in one field may provide theories which will help scientists in other fields to answer other questions.

In formulating hypotheses one should never forget that the proposed tentative solutions may be wrong. Remember that a hypothesis needs to be tested before a final theory or conclusion can be reached. The investigator must keep an open mind while proceeding with his research and look at both negative and positive evidence at each step.

SCOPE AND OBJECTIVES OF THE PROBLEM

One necessary part of identifying clearly the nature of a research activity is the statement on the scope and objectives of the problem under investigation. One of the most frequent mistakes of undergraduates is the selection of research problems that are too large or too difficult for them to handle. Even experienced research analysts often find that they are forced to investigate only one aspect of the larger problem with which they started. The amount of time needed to solve even an apparently simple problem can prove surprising to the inexperienced.

METHODOLOGY

Having found a problem, the student needs to decide on the best method for solving it. If he has prepared a problem analysis report, he will probably have found that people researching similar problems have used a variety of methods to acquire data. In looking at the literature, the student should also be on the lookout for new or unique ways of obtaining and handling data derived from library or field research on his own problem. The written research proposal, therefore, should include a clear, concise description of how the research will be conducted along with copies of any questionnaires or similar documents to be used to support or validate the research.

TENTATIVE OUTLINE OF PAPER

After deciding on the way in which to organize the paper, the student should prepare an outline listing the main topics to be included. Admittedly, the nature of the research report will be modified as investigation and writing proceed, but the student should construct an extensive list of the topics contained in his outline. To this list he should add any pertinent topics that can be easily found in the book *Subject Headings in the Dictionary Catalogs of the Library of Congress*. This will assist him later in finding library materials and in organizing his notes.

In addition to thinking about the organization of the paper, the student needs to think about the kinds of illustrative materials to use in the final version of the report. Planning at the initial stages of research will save much time at a later date. Thus, the tentative outline of the paper should have appended to it a list of illustrative materials as well as a preliminary bibliography to show the sources consulted thus far in the investigation.

SUMMARY

Student research should parallel that of professional geographers. In order to familiarize himself with the ways of scholarship in the field of geography, a student should examine the papers published in the professional journals. In general these papers fall into five main categories: regional synthesis; topical or systematic analysis; evaluation of techniques, theories, or models; literature review; problem solution.

Problem identification is an ongoing activity of professional geographers. Individuals become aware of problems as a consequence of their reading, classroom discussions, attendance at professional meetings, and in a number of other ways. One technique, problem analysis, is a deliberate attempt to unearth research problems through an examination of the literature.

Once a problem has been identified one needs to determine if it is significant, original, and feasible. If it meets all preliminary requirements, the investigator should prepare a research proposal which includes a statement of the problem, a tentative solution (hypothesis), and statements of scope, objectives, and methodology. He should construct a tentative outline of the report, including a list of illustrative materials and a preliminary bibliography.

Problems for Study

1. Prepare a list of research problems that geographers are working on.

2. Prepare a list of journals where you might find papers written by geographers on the following topics: (a) Climatology, (b) Urban Geography, (c) Resource Management, (d) Regional Geography, (e) Methodology, (f) Historical Geography, (g) Geomorphology, (h) Population Geography.

3. Find three examples of research reports that fall into each of the five general categories of papers listed on pages 24 and 25.

4. List a number of research problems that interest you.

5. For three of the problems listed in item 4 suggest several ways to modify the scope of the problem.

6. Prepare a problem analysis report in your general area of interest.

7. Prepare a research proposal on a problem you wish to investigate.

CHAPTER 3
Research Methods

In the previous chapters we have been concerned with the nature of research, with problem identification and delimitation, and with all the preliminary activities necessary before proceeding with data collection. We have seen that research involves the search for undiscovered knowledge; tentative hypotheses are tested to verify the truth or falsity of the ideas contained in them.

The Pattern of Research

The solutions to research problems are arrived at in various ways but usually follow the same general pattern. After defining and delimiting the problem and developing a working hypothesis, the investigator must decide what information he needs and how he is going to get it. At the same time, he needs to decide how he is going to analyze and present his material in final form. Once he has made these decisions he can proceed to seek facts which verify or disprove his hypothesis. In the process of accumulating evidence he may find it necessary to modify his working hypothesis or discard it altogether. If his evidence supports his hypothesis, the investigator may make up his mind that it constitutes the correct solution to his problem and proceed to write up the report.

If you have done any research, you know that such work proceeds slowly, as the following account by Charles Darwin indicates.

When on board H.M.S. "Beagle," as naturalist, I was much struck with certain facts in the distribution of the organic beings inhabiting South America, and in the geological relations of the present to the past inhabitants of that continent. These facts, as will be seen in the latter chapters of this volume, seemed to throw some light on the origin of species—that mystery of mysteries, as it has been called by one of our greatest philosophers. On my return home, it occurred to me, in 1837, that something might perhaps be made out on this question by patiently accumulating and reflecting on all sorts of facts which could possibly have any bearing on it. After five years' work I allowed myself to speculate on the subject, and drew up some short notes; these I enlarged in 1844 into a sketch of the conclusions, which then seemed to me probable: from that period to the present day I have steadily pursued the same object. I hope that I may be excused for entering on these personal details, as I give them to show that I have not been hasty in coming to a decision.

My work is now (1859) nearly finished; but as it will take me many more years to complete it, and as my health is far from strong, I have been urged to publish this Abstract. I have more especially been induced to do this, as Mr. Wallace, who is now studying the natural history of the Malay archipelago, has arrived at almost exactly the same general conclusions that I have on the origin of species. In 1858 he sent me a memoir on this subject, with a request that I would forward it to Sir Charles Lyell, who sent it to the Linnean Society, and it is published in the third volume of the Journal of that society. Sir C. Lyell and Dr. Hooker, who both knew of my work—the latter having read my sketch of 1844—honoured me by thinking it advisable to publish, with Mr. Wallace's excellent memoir, some brief extracts from my manuscripts.

This Abstract, which I now publish, must necessarily be imperfect. I cannot here give references and authorities for my several statements; and I must trust to the reader reposing some confidence in my accuracy. No doubt errors will have crept in, though I hope I have always been cautious in trusting to good authorities alone. I can here give only the general conclusions at which I have arrived, with a few facts in illustration, but which, I hope, in most cases will suffice. No one can feel more sensible than I do of the necessity of hereafter publishing in detail all the facts, with references, on which my conclusions have been grounded; and I hope in a future work to do this. For I am well aware

37

that scarcely a single point is discussed in this volume on which facts cannot be adduced, often apparently leading to conclusions directly opposite to those at which I have arrived. A fair result can be obtained only by fully stating and balancing the facts and arguments on both sides of each question; and this is here impossible.[1]

Primary, Secondary, Tertiary Evidence

In order to satisfy the writing requirements for a research paper you must present your findings in a scholarly manner. Also you must satisfy your reader that you have utilized reliable material and that therefore every detail in your paper is correct. You will do this if you consult authorities and find original or primary sources of information.

Primary evidence from original sources may consist of interviews, diaries, letters, personal observation, government documents, or any other data presented in its original form. *Secondary* evidence consists of research reports which have resulted from the analysis and interpretation of primary sources. Newspaper and magazine articles and popular books may represent *tertiary* evidence if they have been compiled from secondary sources.

Even primary sources need to be examined with care. Autobiographies may show only the good side of a person's life or reflect the author's personal point of view. Climatic data taken with a faulty thermometer is poor evidence of temperature conditions. Quotations taken out of context may present a distorted view of a person's real meaning. In addition, words have different meanings in different eras and in different levels of society. Be sure that you are attributing the correct meaning to the quotations you use.

When secondary sources are used, you need to ask several questions about the author of the reference in question. Was he an authority in the field in which he was writing? Was he in a position to obtain valid data? Was he objective in his approach or was he

[1] Charles Darwin, *The Origin of Species* (Garden City, N.Y.: Doubleday, n.d.), pp. 22–23.

presenting a particular point of view to justify a position already taken?

A tertiary source is better than no source at all, but such sources should be used only if the primary or secondary sources from which they are derived cannot be located.

It should be obvious by now that, in the preparation of a research paper, a student needs to develop a critical sense about the relative value of the evidence which he accumulates. He may be guided by what the authorities say, but he must use his own judgment in the selection and use of materials.

In the pursuit of evidence for your research paper you should act as if you were an attorney preparing a court case. You will need to back up every statement containing original or unusual facts with a citation from an unquestionable authority or produce primary evidence that will prove your point. You don't need to do this if the information is commonly known or universally accepted as true.

Taking Notes

Bibliographic cards should be prepared for each source consulted. The form to be used in the final bibliography should be determined prior to beginning library research and entries should be recorded in this form as the works are examined. The call number, the name of the library in which the book was found, and an evaluation of the relative worth of the item should be noted on the card. Thus, if it becomes necessary to look at the work again, you will be able to find it without difficulty. Such a system also permits you to prepare your bibliography quickly and correctly.

Larger cards should be used for taking content notes. A key word or phrase from the list previously developed (see p. 33) should be placed at the top of each card. As information is acquired, it is entered on a card identified by this key word. Write on one side of the card only, citing the source in sufficient detail so that you can prepare the necessary footnote for each reference. (See pp. 65–66 for discussion of footnotes.) Each card should contain notes on *one* subject only unless the heading clearly states that it is a general set of notes. Single-subject note cards prepared in this way can be sorted and filed in an organized manner, and the information on them is readily available when needed.

In extracting material for a research paper avoid the pitfall of indiscriminate copying of everything pertaining to your topic. It is unnecessary to note the obvious and widely accepted. However, note all facts that are new to you or that are stated uniquely. Notes should include everything that you wish to use in your paper. If you are planning to publish extensive sections of another individual's work, you should obtain permission to do so from both the author and the publisher of the work.

Another pitfall to avoid is that of copying everything in the exact words of the author. Learn to abstract material by putting the author's ideas into your own words. This will save time in the final preparation of your paper. When it is necessary or desirable to quote directly, use the exact words, spelling, and punctuation of the author.

Finally, develop precise, accurate habits of note-taking; it will save you from that last minute rechecking of sources that so many students are forced to do.

In summary, two main points are suggested: standardize your notation scheme prior to beginning research, and abstract material rather than copying whole paragraphs and pages of notes.

Data-Gathering Techniques

Geographers acquire information about problems in two principal ways: through examination of documents and through field observation. For certain kinds of problems both techniques may be necessary. Thus, knowledge of both library and field methods is an essential part of the geographer's bag of tools.

DOCUMENTARY RESEARCH

In the broad sense, documentary research consists of locating and extracting information from written works, including unpublished records of governmental and private agencies, personal diaries, newspapers, journals, and books. Such materials may be located in a number of different places and a part of the scholar's job is to identify those places which have information of value to him. For the undergraduate student, most of the information sources which he will use are in his college library. In this guide our primary concern is to provide the reader with assistance in the identification and utilization of library materials.

Almost every library has material useful to you in your research on the topic you have chosen to investigate. Your immediate problem is to find these materials and to extract information from them. How do you proceed?

The first thing to do, if you have not already done so, is to become familiar with the library where you will be working. If the library has a handbook, get a copy and read it. If the library conducts tours for visitors, join one. If no such opportunities exist, take a tour of your own.

What should you look for in the library? First of all, where is the card catalog? What is listed in it? Is there a separate catalog for serials (periodicals)? For government documents? Special collections? Is the Library of Congress or Dewey decimal system used in classifying books? Where is the reference section? Is it mainly bibliographical or does it contain content materials? Is there a reference librarian available to help you locate information?

How do you obtain books? Determine whether you have access to the stacks and, if so, make a quick tour through them. Make a note of the places where books related to your topic are shelved. Where are the periodicals kept? Government documents? Microfilm? Special collections? Is there a reserve reading room in which materials utilized in courses are kept?

After completing your quick tour of the library you are now

ready to start doing your research. In the tour you will have seen two major classes of material useful to you—bibliographical tools and content materials—from which to obtain facts and ideas. Which of these should you utilize first? The answer to that question depends on the knowledge that you have about your research topic. If the topic is relatively new to you, consult a general source, such as an encyclopedia or a textbook, for additional information on your subject. Consult A *Basic Geographical Library*, a bibliography published by the Association of American Geographers, for the appropriate secondary source.[2]

In consulting a secondary source (such as the encyclopedia or textbook) you may find references to additional materials that are footnoted or contain rather complete bibliographies on your subject. However, you may not make such a fortuitous discovery, and the job of uncovering information is not always so easy.

CARD CATALOG

The card catalog is the key to the identification and location of source materials available in the particular library in which it is located. For any scholar knowledge of the card catalog and its use is essential. Almost all libraries use a system consisting of a continuous alphabet of cards which give the reader information about the book and an indication of its location in the library.

Each book is assigned a number or a letter which is determined by the general subject of the book. To this number or letter are added other letters and numbers which further identify this particular book. This combination of numbers and letters imprinted on the book and on the upper left corner of each catalog card is known as the *call number* of the book. It is the guide by which the book is located on the shelves of a library, whether the stacks are open or closed. Most open-stack libraries post charts showing the location of groups of books with common call numbers.

[2] Association of American Geographers Commission on College Geography, A *Basic Geographical Library: A Selected and Annotated Book List for American Colleges,* Publication No. 2, compiled and edited by Martha Church, Robert E. Huke, and Wilbur Zelinsky. Washington: 1966. Also 1970.

Classification Systems

There are two principal classification systems currently in use in libraries of this country. The system in use in many public libraries and in some older university libraries is the Dewey decimal system, devised by Melvil Dewey of Columbia University. In this system categories have fixed numbers; each author is identified by a fixed letter (almost always the initial of his last name); and each book is identified by an additional number.

The ten main groups of the Dewey decimal system are:

000–099	General Works	500–599	Pure Science
100–199	Philosophy	600–699	Useful Arts
200–299	Religion	700–799	Fine Arts
300–399	Social Sciences	800–899	Literature
400–499	Philology	900–999	History

The Library of Congress classification system is used by most large research libraries. In this system capital letters are used to identify major classes of books, with numbers used after the letters to designate subdivisions by subject, by form, or by geographical location. A letter (generally the initial of the author) and a number further identify each book. The year of publication may also be added to complete the identification of the book. The main classes of books are assigned the following letters.

A	General Works	J	Political Science
B	Philosophy and Religion	K	Law
C	History and Auxiliary Sciences	L	Education
		M	Music
D	History (except American)	N	Art
		P	Language and Literature
E–F	History, American	Q	Science (General)
G	Geography and Anthropology		QA Mathematics
			QB Astronomy
H	Social Sciences		QC Physics and Meteorology
	HB–HJ Economics		
	HM–HX Sociology		QD Chemistry

QE Geology	T	Technology
QH–QR Biological Sciences	U	Military Science
	V	Naval Science
R Medicine	Z	Bibliography
S Agriculture		

The letters, I, O, W, X, and Y are not presently utilized.

Regardless of which classification system is used, the books used by geographers are scattered throughout the library. Geography students should familiarize themselves with the kinds of material to be found under the categories listed above. These represent most of the subclasses of books containing material of a geographical nature.

Types of Cards in the Catalog

Most books are listed at least three times in the card catalog, under author, title, and subject.

AUTHOR: This card is the main entry in the catalog. Instead of a person, the author may be a corporate body (American Geographical Society) or a government agency (U.S. Bureau of the Census). In the case of periodicals and some general works, the main entry may be the title card.

TITLE: The title is generally typed in black above the author's name on the title card. It is often easier to locate a book, especially those of the federal government, by title rather than by the corporate author.

SUBJECT: In preparing the list of topic headings related to your problem you made use of the volume entitled *Subject Headings Used in the Dictionary Catalogs of the Library of Congress* (see p. 33). This book lists in alphabetical order the subjects that are used as headings and shows the applicable cross-references. Books containing information on specific subjects are listed under these headings. Subject headings may be typed in red or in capital letters.

The sample cards on page 45 illustrate the three ways a book may be listed in the card catalog.

G
1525
D8

Durrenberger, Robert W
    Patterns on the land; geographical, historical, and poli-
tical maps of California ₍by₎ Robert W. Durrenberger.
Harold Schwarm, graphic designer; Donald Ryan, cartographer; Leonard Pitt, historical consultant. ₍3d ed.₎ Woodland Hills, Calif., Aegeus Pub. Co. ₍1965₎

    vi, 109 p. illus. (part col.) maps (part col.) 32 cm.

    1. California—Maps. 2. California—Econ. condit.—Maps. 3. Cali-
fornia—Historical geography—Maps. ɪ. Title. ɪɪ. Title: Geographical, historical, and political maps of California.

Patterns on the land.

G
1525
D8

Durrenberger, Robert W
    Patterns on the land; geographical, historical, and poli-
tical maps of California ₍by₎ Robert W. Durrenberger.
Harold Schwarm, graphic designer; Donald Ryan, cartographer; Leonard Pitt, historical consultant. ₍3d ed.₎ Woodland Hills, Calif., Aegeus Pub. Co. ₍1965₎

    vi, 109 p. illus. (part col.) maps (part col.) 32 cm.

    1. California—Maps. 2. California—Econ. condit.—Maps. 3. Cali-
fornia—Historical geography—Maps. ɪ. Title. ɪɪ. Title: Geographical, historical, and political maps of California.

California--Maps

G
1525
D8

Durrenberger, Robert W
    Patterns on the land; geographical, historical, and poli-
tical maps of California ₍by₎ Robert W. Durrenberger.
Harold Schwarm, graphic designer; Donald Ryan, cartographer; Leonard Pitt, historical consultant. ₍3d ed.₎ Woodland Hills, Calif., Aegeus Pub. Co. ₍1965₎

    vi, 109 p. illus. (part col.) maps (part col.) 32 cm.

    1. California—Maps. 2. California—Econ. condit.—Maps. 3. Cali-
fornia—Historical geography—Maps. ɪ. Title. ɪɪ. Title: Geographical, historical, and political maps of California.

G1525 .D8   1965   912.794   Map 66–409

Library of Congress   ₍2₎

45

Interpreting the Catalog Card

The catalog card provides an excellent approach to the content of books. In many instances, works may be selected or rejected by a skillful interpretation of the data on the cards.

Note the information given on the card shown below.

1. *Author's (or editor's, compiler's, etc.) name and date of birth.*
2. *Title of book.*
3. *Imprint:* Place and date of publication and name of publisher.
4. *Collation:* Description of the physical properties of a book, such as number of volumes, pages, size, illustrations, maps, etc.
5. *Tracings:* These are a record of all listings, other than by author, under which the book may be found in the catalog. There is a subject tracing given for each important subject treated in the book. By looking under these subject headings in the catalog, the reader is able to find what other books this library has on the same subject. Additional tracings for joint author, title, or series are also indicated. The sample card shows that there are

California--Maps

G
1525
D8

Durrenberger, Robert W
Patterns on the land; geographical, historical, and political maps of California ₍by₎ Robert W. Durrenberger. Harold Schwarm, graphic designer; Donald Ryan, cartographer; Leonard Pitt, historical consultant. ₍3d ed.₎ Woodland Hills, Calif., Aegeus Pub. Co. ₍1965₎

vi, 109 p. illus. (part col.) maps (part col.) 32 cm.

1. California—Maps. 2. California—Econ. condit.—Maps. 3. California—Historical geography—Maps. I. Title. II. Title: Geographical, historical, and political maps of California.

G1525 .D8   1965   912.794   Map 66–409

Library of Congress   ₍2₎

four cards in the catalog representing this book: one for the first author, one for the joint author, one for the title, and one for the subject.

6. *The call number:* The call number is a combination of the class number (G 1525), which represents the major subject of the book, and the book number (D 8), which represents the author's name. Books are shelved in the stacks as if there were a decimal point after the first numeral of the book number, i.e., books with these numbers would shelve in this order.

G	G	G
1525	1525	1525
D76	D8	D85

Additional symbols are added where necessary to differentiate titles, edition dates, and volume and copy numbers.

Symbols indicating that books are in a special location instead of their regular class order on the shelves are placed above the rest of the call number (f).

7. *Subject heading:* A word or phrase is used on the cards in the catalog to bring together material dealing with the same subject.

Topics or themes being investigated are often thought of by the reader in different words or phrases from the ones used in the card catalog. The student who does not find a subject for which he is looking may find, instead, a "See" reference card, or he may consult the Library of Congress subject headings list.

The Arrangement of Cards in the Catalog

Because of the tremendous number of cards in the catalogs of libraries rather complex rules have been devised that in some cases develop a logical rather than an alphabetical arrangement. Although some variation in these rules occurs from one library to another, knowledge of some of the more general rules is helpful.

1. Cards are filed letter by letter to the end of a word, and then word by word. (For example, "American art" is filed before "Americana," and "New York" is filed before "Newark.")

2. Abbreviations, such as Mr., Dr., U.S., St., are arranged as if spelled out in full in the language of the entry.

> St. Lawrence = Saint Lawrence
> S. Lorenzo = San Lorenzo

3. Numbers are filed as though written out in the language of the entry.

> 8 days = Eight days
> 8 jours = Huit jours

4. All variations of the proper name beginning with "Mac" (Mc, M') are filed as "Mac" and are interfiled with all other words beginning with "mac."

5. The articles *a*, *an*, and *the* and their foreign equivalents are disregarded in filing only if they are the first word of a title or entry.

6. Subject cards are filed systematically according to the punctuation in the heading. The simple subject comes first, then subjects arranged alphabetically by subdivision, then subjects with parenthetical phrases, those with inverted modifiers, and finally phrase subject headings. (The punctuation order is dash, parenthesis, comma.) Within each subject heading, cards are filed alphabetically by author.

   The following example illustrates the order of filing modified subject entries.

> Cookery
> Cookery—Dictionaries
> Cookery—History
> Cookery (Fish)
> Cookery (Mushrooms)
> Cookery, American
> Cookery, West Indian
> Cookery for camps

7. Where the same word may be used for a person, a place, a thing, and the first word by a title, the entries are filed in that order, as shown in the following example.

Orange, Roger L.	(Personal name, as author)
Orange, Roger L.	(Personal name, as subject)
Orange, N.J. University	(Corporate author)
Orange, N.J. University	(Corporate entry, as subject)
Orange	(Thing, as subject entry)
Orange—Cultivation	(Thing, subdivided subject)
Orange, Navel	(Thing, inverted modifier)
Orange juice	(Thing, phrase heading)
The orange and I.	(First word of title)

8. Cards for an author's books are filed under his name, alphabetically by title, except that cards for his complete works precede the individual titles.

   Cards for books *about* the author are filed *after* all the cards for his works. Title cards for books which begin with the author's name are filed last.

Keats, John	(Author card)
Keats, John	(Subject card)
Keats, Shelley, Byron, Hunt	
and their circles.	(Title card)

9. Corporate entries follow the same general rules, but appear more complex. A work published by a governmental body is entered under the name of the agency of which it is a part. There are many cards in the catalog under the various states, cities, counties, and countries. As with personal authors, there are books *by* the various agencies, which are followed by books *about* each one.

   Then come the subject cards for the country, state, etc., as a whole, followed by title cards for books whose titles begin with the name of the country, etc., interfiled with cards for unofficial organizations as author. The example which follows illustrates the main essentials of the sequence.

U.S. Congress	(As author)
U.S. Congress	(As subject)

49

U.S. Weather Bureau   (As author)
U.S. Weather Bureau   (As subject)

At the end of the alphabetical sequence of U.S. government agencies are found the subject cards for books about the United States (country) as a whole.

U.S.—Description and travel
U.S.—Economic conditions
U.S.—History

Many subject headings are subdivided historically, and these are arranged in chronological order.

U.S.—History—Colonial period
U.S.—History—Revolution
U.S.—History—Civil War

Title cards and unofficial organizations file after all the subject cards.

The United States as seen by
  Latin American writers   (Title card)
United States Steel Corporation   (As author)

Many useful reference works are listed in Part II of this guide. Learn where the commonly used reference materials are shelved in the library and ask the librarians to help you in locating those items you cannot find. Their job is to assist you in the use of library materials. Familiarity with the library facilities will save you many hours in the search for information.

Field Techniques

The facts that geographers use in their research have accumulated as a result of the exploratory and interrogative efforts of many individuals. "Armchair geographers" may make use of data gathered by others, but exponents of geographical field work maintain that unless one initiates and gathers the data himself he is not really doing research. The ridiculous nature of this statement should be obvious to all; there would be no historical research if it were true, and any study of the simultaneous distribution of any element of the landscape over much of the earth's surface impossible.

However, in the past, there has been too great a tendency for geographers to depend upon scientists in other disciplines to generate data for them to analyze. There is a happy medium in which the investigator does field work to verify data gathered by someone else or to develop additional facts or information. Each problem presents different kinds and degrees of difficulty for the investigator, and procedures to solve each problem need to be developed on an individual basis. Thus, it is not possible to produce a field manual covering all possible techniques which a geographer might use in field study. Rather, in this guide we will outline a few commonsense rules to follow in gathering data in the field.

PREPARATION FOR FIELD WORK

One does not just go out to the problem area and begin to accumulate data. Rather, one must decide precisely what data is needed, how it is going to be gathered, and how it is going to be recorded. Field work is not a separate part of research, isolated from other aspects of problem-solving and accomplished without library research. As a matter of fact, detailed field study should commence only after the investigator has exploited all other sources of information relating to his problem.

Naturally, a reconnaissance trip to the problem area is useful in getting an overall view of the kinds of information available and to help develop a research plan. But, first of all, a careful search should be made of previous work done in the area. Too often, a field investigator, after completing a difficult piece of work, has been confronted with a document containing precisely the kind of data he has worked so hard to obtain in the field.

And, unless one has familiarized himself with the literature dealing with his problem and has given a great deal of thought to the kinds of information he is seeking to obtain by field work, a good deal of time may be lost. It must be remembered that field study, although pleasurable, is also time-consuming. The decisions on the kinds of information needed and the procedures for acquiring and recording data need to be worked out prior to commencing field study.

INTERVIEWS AND QUESTIONNAIRES

Much useful information is obtained in the field by communicating with people either orally or in writing. While methods of interviewing people or of designing questionnaires cannot be covered adequately within the scope of this book, a few general ideas and specific rules of thumb will be suggested.

First of all, it should be pointed out that the process of getting information from another person is susceptible to both deliberate and unconscious abuse. Even the objective investigator may err in the way he constructs or asks his questions or in the way he interprets the information received. The person doing the interviewing invariably influences the quality of the information collected. For this reason, it is imperative that students using the interview method for the first time should attempt to remain as objective and neutral as possible and not try to influence the person being interviewed. In other words, the interviewer must probe for responses without letting his own attitudes and opinions interfere with the process of data collection.

Questionnaires should be pretested under realistic conditions before being used; often retests are needed to check revisions made after the initial trials. The information desired should be defined carefully and precisely. It is necessary to confine the scope of any questionnaire by establishing priorities for the things that are really needed over those that would be "interesting to know." Only after it has been decided exactly what kinds of information are essential, can questions be formulated and structured to obtain the needed data. Questions should be neutral in their wording and phrased so that people will be willing and able to answer them. Sentences should be short and the wording clear to prevent misunderstanding the intent. It is often advantageous to explore a significant point by several different but related questions.

You need to construct a rather precise timetable for the completion of any data-gathering procedure with sufficient allowance for any follow-up work necessary to verify or gather additional information. You must also consider the time of the individuals upon whom you are dependent for information. Find out when interviews may be accomplished most efficiently and expeditiously so that the greatest amount of information may be acquired in the least time. Prior to an interview, mail a set of questions to the interviewee so that he will be prepared to answer your questions fully and accurately. At the time of the interview arrangements should be made for any additional interviews that might be needed.

Although specialized experience in interviewing and in drafting questionnaires is desirable, there is no substitute for common sense and courtesy in eliciting good responses. The right approach and the right questions will generally get you the right answers to help you solve your problem.

FIELD-MAPPING

If your problem involves the gathering of data that can be mapped in the field, find or construct a base map at a scale allowing for inclusion of all the data that you wish to record. U. S. Geological Survey topographic sheets make excellent base maps for many purposes. Where more than one feature of the landscape is to be plotted, plastic overlaps may be used. Where available, aerial photographs also make excellent bases on which to record data. Again, use of plastic overlays is recommended.

It is impossible to record all of reality on a base map or even on an aerial photo. In some cases, as with cultural or political boundaries, the lines exist only in the minds of men. In other cases, the number of individuals to be observed and recorded are too many and techniques to sample the distributions must be utilized. One of the most commonly used sampling techniques is the transect in which the investigator selects a line extending across the problem area and records the data observed along the line. In other cases, he may simply take observations at random points located within his study area and analyze this data with the notion that random samples are representative of the data pool as a whole.

OTHER FIELD ACTIVITIES

In certain types of studies, such as those involving atmospheric research, geographers often must establish a network of data-collecting stations on their own. The selection of representative sites and correct exposure then become extremely important if the data collected are to have significance. A well-organized study on a significant problem becomes worthless if the conclusions are based on inadequate or inaccurate data.

Upon return from field work each day, one should organize, evaluate, and summarize the data gathered because it is easy to misplace data or forget some significant event or fact by waiting too long to record and evaluate it. Such summation of the day's activities also prepares one for the next day's work and makes it possible to clarify quickly points at question. Follow-up visits to the research area are also more meaningful if each day's activities have been carefully and completely recorded.

SUMMARY

The preparation of a research paper involves five kinds of activities: problem identification, planning, data-gathering, data analysis and interpretation, and report-writing. Chapter 3 deals principally with data-gathering techniques involving library and field methods. It discusses the use and evaluation of evidence and ways of recording the data which have been gathered.

For most students the principal source of evidence will be their college library. They need to become familiar with the kinds of reference materials available to them and to know how to make use of them. In addition, they need to know how to find articles appearing in periodicals and how to use government documents. They should familiarize themselves with the card catalog and with the location of books in their library.

For some problems, data can be gathered only by field work. One must decide what data is needed, how it is going to be gathered, and how it is going to be recorded before ever venturing into the field. Because of the time-consuming nature of field work great thought must be given to planning every detail of the field excursion. If one is to use techniques involving interviews or questionnaires, great care must be taken to avoid bias in either questions or answers. A diary should accurately summarize all significant research activities of each day.

Problems for Study

1. Discuss the kinds of evidence that geographers use in their research. List examples of (a) primary, (b) secondary, and (c) tertiary evidence.

2. Prepare bibliographic cards for the following types of references: (a) book, (b) periodical article, (c) government agency publication, (d) U.S. Congress publication, (e) unpublished document, (f) interview.

3. Prepare a content note card on a topic of your own choosing. Make use of actual reference materials in preparing it and include both abstracted and quoted materials.

4. You have just been chosen to head up a new department of geography at a college. The librarian has asked you to prepare a list of the

most significant reference items needed by geographers. What items would you include?

5. Do the same for geographical journals.

6. You have been awarded an ONR grant to work on a field problem in a foreign country. Draw up a plan for completing the necessary field work.

CHAPTER 4

Preparing
the Manuscript

One of the major points made in earlier sections of this guide is the importance of keeping in mind the nature of the audience toward which your paper is directed. In almost all cases student papers are directed toward satisfying the demands of a professional geographer. That is why the material on the following pages closely parallels the editorial policy statement of the *Annals,* Association of American Geographers (Appendix II).

This chapter outlines steps to be followed in completing your manuscript for submission to your instructor. The guidelines contained herein represent suggestions only. Each instructor may wish to modify the final form in which your manuscript is prepared. Check with him to see if he wishes you to follow these instructions rigidly.

Planning

Breaching the barrier between the data-gathering process and the writing process is a difficult task for most people. This is because they do not realize that writing involves three major steps: planning, preparing a rough draft, and preparing the final manuscript. If there is no well-thought-out plan for the paper, preparation of the rough draft appears to be an insurmountable obstacle and the writing of an opening statement an impossible task. Students soon learn that more time spent in the planning stage will mean more time saved in the writing and revision stages.

One of the most significant steps in the planning process is the choice of the basic structure about which you fashion your report. As you have been told previously, there is no single way to organize a research paper in geography. Rather, the form that your paper takes will be determined by the nature of the report itself as well as by other considerations. In Chapter 2 it was suggested that student research reports in geography fall into five broad categories: regional synthesis; topical or systematic analysis; evaluation of techniques, theories, or models; literature review; and problem solution. Additionally, within these categories are a variety of possible types of organization. Spend considerable time and thought on the preparation of your working outline; for, even though you have chosen a significant problem and gathered adequate data, it will do you little good if you are unable to organize your material in an acceptable form.

Earlier, you were advised to prepare an outline of your paper prior to starting the search for information. As you acquired greater knowledge of your problem, your original ideas on the arrangement of various topics in your paper may have changed. Now that you have finished gathering data and doing field research, it is time to take a look at the original outline to see that you understand the purpose of your paper and the nature of the audience for whom it is intended.

In addition, answers to the following questions will clarify the approach to the organization of your material. What is the central theme of your paper? What is your attitude toward the central theme? Are you approaching the problem objectively or with a bias? What tentative conclusions have you reached concerning the problem?

Now read through your notes once more and separate them

into piles according to the topic headings that you have chosen. Is there a natural order into which these topics fall? Are any of these topics related so that they would fit together into larger topics? Do you see any kind of natural order to the material that you have accumulated? If you are unable to discern a natural order to the topics on which you have taken notes, look at reports in the same category as yours which appear in the major journals and use them as a guide in preparing your working outline.

One of the first principles in writing is to understand the structure of the paper you are preparing and to build paragraphs which follow and fill out the structure. A good writer will, in part, follow his outline and, in part, deviate from it as he finds it necessary. But, prior to writing even the introductory paragraph, it is a good idea to establish a hierarchy of headings for the topics and subtopics on which you plan to write.

Also before writing there should be some planning on the nature and number of tables and other illustrative and reference materials to be used. Many times students reach the stage of typing their final draft before thinking about including such materials. Illustrative and reference materials not only add to a report but simplify the task of the writer; it is an easy matter to amplify material by using a map, graph, or photograph. Thus, a tentative list of tables, graphs, maps, photos, and other reference materials should be prepared before any writing is done.

Planning should also consider the time element, for almost everything that is written is governed by a completion date. You need to establish deadlines and stick to them. Sufficient time must be included in your timetable to revise your paper, to complete your illustrations, and to edit and make corrections on the final manuscript. Developing a sound plan for completion of your research report is one of the most significant parts of the research process. Time spent in planning will result in time saved in preparing the manuscript and in a product of higher quality.

Writing the Report

Writing may be compared to many other kinds of human activities. A sculptor working in clay to form a bust of a famous man has raw material and a sketch to work with. The relative success of his efforts will depend on the quality of three aspects of his work; he must have chosen his raw material carefully, developed a good sketch to work from, and manipulated his raw material skillfully in order to achieve the desired results. A good research paper is put together in much the same way as any other art form. A writer also fashions his work from raw material, his notes, according to a sketch, or outline, previously established. The degree of skill he exhibits in fashioning his paper will in large part determine the quality of his paper. Moreover, like the sculptor, he must have good raw material and a good sketch, or outline, in order to fully utilize his skills as an artist.

The importance of diligent, efficient research will become apparent when you attempt to fill in the details surrounding the skeleton outline that you have created. If you have large gaps in the information necessary to complete your paper or if your notes were taken in a loose and sloppy fashion, you will find it difficult to write a satisfactory research paper. Effective research and a systematic method of recording data are basic requirements for successful completion of a paper. High quality notes and good ideas represent first-rate raw materials from which to form your paper.

A good outline is as important to the writer as a good sketch is to the sculptor. If the report you are preparing is a short one there may be no need to subdivide it into topics, but for most research reports in geography the material should be organized into sections, topics, and subtopics. Care should be taken to keep topics of equal significance at the same level so that important ideas are not placed in a position subordinate to less important ideas, and vice versa. Effective arrangement and development of facts and ideas is one of the essential features of a good research paper.

Even though all of the students in an art class start with the same materials and the same model, their finished products are different. Each person has his own way of expressing himself—his own style. Good style in writing consists of an effective choice of words and of good judgment in the development of sentence patterns and paragraphs so that the writer satisfies the purpose of his paper. A beginning writer should turn away from embellishments and devices that clutter the page and obscure the ideas; he should strive to phrase his thoughts in clear, concise sentences and paragraphs.

Effective writing is natural, simple, and economical. It involves saying what you mean in as few words as necessary and in language that is understandable. Writing too much is a common fault. If you have material that does not belong in your paper, throw it out or save it for use another day. Say only as much as you need to say on any given topic, not as much as you can say; for the ease with which a reader can understand your thoughts is inversely related to the number of words used to express them. Use commonly accepted words rather than jargon or colloquial phrases, for you are preparing a research report—not a newspaper article.

Despite the previous admonitions, research papers need not be dull, heavy reading. They can be written so that ideas are interesting and easily understood. Consider rhythm and motion of words, which are important aspects of style. By varying your choice of words and sentence and paragraph structure, you can avoid monotony. Also try mixing long and short, simple, compound, and complex sentences. The rate at which you develop your thoughts is also important. In climbing a mountain trail you do not move upward at a uniform rate. Where the trail is difficult, you move slowly and cautiously, choosing your steps with care so that you do not stumble or fall. In difficult parts of your paper you also should move slowly, making use of tables and illustrative materials, choosing your words with care, and, where necessary, retracing your steps so that the reader will comprehend what you have said before moving ahead to the next idea.

BEGINNING TO WRITE

Once you are satisfied with your plan, begin to write. If you have chosen your topic well, have done extensive research, and have a well-conceived plan, the task should be relatively easy. But, for most people the question is, "How to begin?" The answer is that there is no one way to begin a research paper. Remember that unless you get the attention of the reader immediately, he may never finish your paper. For him, your first words may also be your last words. Therefore, the importance of a good beginning cannot be overemphasized. Most readers of scholarly articles are busy people. They must make a decision on the basis of the first several sentences whether to read further or to discard the article. Rightly or wrongly, writers are frequently judged on the quality of their opening paragraphs. Poor writing in the middle of a paper may escape notice—opening sentences stand in the spotlight.

One sound way to begin a research paper is to present background material necessary to understand the problem of the study. Another is to state the problem and the purpose of the research on which you are reporting. You might benefit from an examination of introductory statements which appear in textbooks and in articles in professional journals. You will find, in addition to the two types of introductions cited above, other devices also used, such as quotations, questions, and forecasts based on the findings in the report. You need to consider the nature of your audience and of your report in phrasing the introductory statement; the statement should also reflect your own personal style and not be out of character with the material which follows.

With your outline before you and your notes separated by topics, begin to write your report. Review briefly your notes on the topic you have chosen to start with. Once starting to write on a given topic, do not attempt at this point to polish your material. When using quoted or paraphrased material from your notes, do not recopy these statements but clip the notes to your rough draft in the appropriate places. The principal purpose of preparing a rough draft in this way is to permit you to concentrate on one part of your paper at a time. Your mind will operate more quickly than your pen. If you write on one topic at a time, you may capture and express good ideas while fresh in your mind. However, never take your eyes off the central theme of your paper or of the audience to which it is directed. It is *your* paper and your research efforts should form the substance of the paper; your ideas and your plan of organization should come through clearly; and the paper should reflect your own unique way of phrasing the facts and ideas contained therein.

If you have developed a good outline, each section of your paper will deal with a natural subdivision of the major theme. In the first draft your principal concern should be with the internal unity and coherence of each section. Transition from one section to the next can be achieved best in later revisions. However, at all times you need to consider the relationship of all the parts to the whole.

Paragraphs represent further logical divisions of thought within each section. In general, each paragraph should develop one main idea completely and at some length. Paragraph lengths may need to be modified to suit the format of the journal in which an article is published or to break the visual monotony of long unbroken blocks of type. Long paragraphs in a journal such as *Science* would overwhelm the reader whereas the same paragraph would be perfectly acceptable in the *Geographical Review*.

To achieve unity and coherence within any given section of your paper, you should make the ideas flow from beginning to end. Thus, good beginning and ending sentences in paragraphs and sections of a paper are very important. The relationship of each paragraph to the ones that precede and follow it, and to the section as a whole, needs to be made clear. This is generally accomplished by references to material previously or subsequently discussed, by topic sentences, and by transitional sentences.

Without coherence within paragraphs and sections the whole paper will fall apart. There is nothing more disconcerting to the reader than to find out that he has been led in a circle back to a point previously discussed. It is extremely annoying to be tantalized with a few facts and ideas here and there and then find more details in another part of the paper. Such scattering of facts and ideas leaves a confused and contradictory impression in the mind of the reader.

Just as good sentences are needed at the end of paragraphs and sections of your paper so too do you need a good ending to your essay. It is the reader's last hope of comprehending the purpose of your paper if your previous statements have been obscure. His total impression of the quality of your paper may be largely based on the quality of your concluding remarks. Thus, you need to give due consideration to what you say in them.

Good endings may take a number of different forms. They may: (1) summarize your work, (2) contain a prediction based on your findings, (3) analyze the significance of your findings, (4) contain recommendations for future courses of action based on your findings, (5) combine two or more of the preceding.

Your concluding remarks should in no way depart from the style and tone of the rest of your paper but should form a harmonious part of the total. They should be truthful and should represent an effective ending to what you have had to say. A good conclusion leaves the reader satisfied that you have accomplished the purpose of your paper; it should represent your very best writing effort.

Footnotes and Footnoting

Footnotes used in scholarly works are of two kinds: (1) explanatory and (2) reference citations.

Explanatory footnotes contain information the author wishes the reader to have, but which would interfere with the flow of thought if located within the text. Many individuals, particularly in the natural sciences, object to the use of explanatory footnotes, but such notes are widely used in the social sciences and humanities. If the idea can be worked into the text, it is generally better to do so. If the idea is unimportant, omit it.

Reference citations indicate the source of direct and indirect quotations which are included in the body of your paper. Those facts which are generally known are considered public property, and you are free to use them without citation. But any fact or idea obtained from another source must be credited to the original author.

In the section on note-taking (pp. 39–40) you were advised to paraphrase and and digest much of the material which you found in doing library research. Occasionally, you may need to quote directly from a source to prove a particular point, but quotations should be kept short, and they must merge smoothly into the text. They should be used when the language of the cited authority is exceptionally appropriate or when you are using an original source to verify a statement you are making.

Remember that footnotes are intended to aid the reader in understanding the thought processes developed in your paper

65

and in seeking out additional information about topics discussed. A beginning student in any academic field is best advised to footnote every unusual fact or idea which the reader might question or wish to examine further.

The appropriate forms to be used in footnotes are to be found in the *Annals* style sheet on pages 114–15. See page 79 for final preparation of footnotes. For additional examples, examine articles in the *Annals,* Association of American Geographers. Different journals have different systems. So, if you are planning on publication, be sure to check with the journal for its preferred style.

Bibliography

A bibliography should include all of the source materials which you consulted in the preparation of your text. As with footnotes, there is no universally adopted form in which bibliographic entries are prepared. However, the form you use should be simple, logical, and consistent. See page 79 for instructions on final preparation of bibliographies.

The bibliography contains the same kinds of information, in more detail, as do footnotes, but serves a different purpose. It should enable anyone, with a minimum amount of difficulty, to find in a library or bookstore each item listed. Also, it should give the reader a good idea of the nature of the source material from which you have constructed your paper. An informed reader can gain some insight into the value of your paper by checking the bibliography to see if you have consulted recognized authorities and used reliable primary and secondary sources.

Ordinarily, your bibliography will include all of your references in one alphabetical list, but for longer papers, theses, or doctoral dissertations, you may wish to prepare a classified bibliography. In such bibliographies the materials may be divided into a number of subclasses. One form commonly used lists books, periodicals, newspapers, and public documents separately.

Any well-made bibliography gives the following information on book entries.

1. Name of the author in the form in which it appears on the title page, in inverted order for alphabetizing
2. Title and subtitle
3. The edition if other than the first
4. Place of publication
5. Publisher
6. Date of first publication (copyright date)

It gives the following information on articles.

1. Name of the author, again inverted
2. Title of the article
3. Title of the periodical
4. Volume and, if there is one, number
5. Month and year of publication
6. Page numbers (inclusive)

Examples

Thompson, John, "The Settlement Geography of the Sacramento–San Joaquin Delta," *Bulletin, California Council of Geography Teachers*, 4 (1956–57), 3–7.

———, *The Settlement Geography of the Sacramento–San Joaquin Delta, California,* unpublished doctoral dissertation in Geography, Stanford, 1958.

———, *The Settlement Geography of the San Gorgonio Pass Area,* unpublished master's thesis in Geography, University of California, Berkeley, 1951.

Thompson, Rolland, *The Railroad Boom in the Los Angeles Area, 1886–1890,* unpublished master's thesis in History, Claremont Graduate School, 1942.

Thornton, J. Quinn, *The California Tragedy*. Oakland: Gillick Press, 1945.

Thurman, A. Odell, *The Negro in California Before 1890,* unpublished master's thesis in History, University of the Pacific, 1945.

Treadwell, Edward F., *The Cattle King.* New York: Macmillan, 1931.

Truman, Ben C., *Semi-Tropical California.* San Francisco: A. L. Bancroft and Co., 1874.

Ulibarri, R. O., *American Interest in the Spanish-Mexican Southwest, 1803–1848,* unpublished doctoral dissertation, University of Utah, 1963.

Ulsh, Emily, *Doctor John Marsh, California Pioneer, 1836–1856,* unpublished master's thesis in History, University of California, Berkeley, 1924.

Underhill, Reuben L., *From Cowhides to Golden Fleece.* Stanford: Stanford University Press, 1939.

U. S. Congress, 31st House of Representatives, 1st session, *California and New Mexico.* Washington: 1850.

U. S. Congress, 31st House of Representatives, *Report on California.* House Document 59 by Thomas B. King, Washington: Gideon & Co., 1850.

U. S. Congress, 35th House of Representatives, 1st session. *Pacific Wagon Roads.* Executive Document No. 108 by Albert H. Cambell, Washington: 1859.

U. S. Department of Agriculture, "Reclamation of Swamp and Overflowed Lands in California," in *Report of the Commissioner of Agriculture, 1872.* Washington: 1873.

U. S. President, *Message of the President of the United States to the Two Houses of Congress.* Washington: Wendell and Van Benthuyson, 1848.

Revision

After completing the rough draft of your report, go through it to check section and topic headings and to construct a tentative table of contents. This permits you to look once again at the structure of your report and helps you to see where revisions or additions are needed. By now, you may be fairly tired of working on your report. Most people benefit from laying their work aside for a short time before attempting to revise it. They are then able to reevaluate their work from a fresh viewpoint. When you are ready to go back to work, read through your paper completely. Look again at the purpose and scope of your report and the plan which you developed to fulfill your purpose. Ask yourself the following questions.

1. Is the purpose of my paper perfectly clear?

2. Have I fulfilled the purpose?

3. Is the plan that I have developed to fulfill my purpose a sound one?

4. Have I followed my plan?

5. If not, should I alter my plan?

6. Does my introduction adequately prepare the reader for what follows?

7. Have I given appropriate emphasis to the important ideas?

8. Have I overemphasized secondary ideas?

9. Is there extraneous material?

10. Are there gaps which need to be filled?

11. Does each paragraph and each section develop one idea fully and completely?

12. Have I linked one section to another and shown the relationship of each part to the whole?

13. Have I utilized variety in language, in sentence and paragraph structure, and in the ways in which I have developed ideas?

14. Does my conclusion terminate and round off the paper?

15. Have I told my story in as few words as possible?

Now commence to modify your paper. Just as a sculptor adds or removes clay in his search for perfection in his completed work so too must a writer change, transpose, add, and cut material from his paper in order to achieve a satisfactory finished product. In doing this, think back to what was previously said about structure and style. Can everyone understand what you mean? Is the logic of your ideas evident? Is their sequence valid?

If the important ideas are not logically organized and clearly evident, consider ways to restructure your material and to rephrase your statements. Do not be afraid to eliminate secondary ideas even though they may be personal favorites of yours. Keep your eyes on the main goal—that of fulfilling the purpose of your paper. Each sentence, each paragraph, now needs to be considered in its relation to those which precede and follow it. Within sentences, words or phrases that are related to each other need to be brought into close proximity with each other. The same holds true of related sentences within paragraphs and of paragraphs within sections. Now is the time to consult a dictionary and a thesaurus to check meanings and spellings and to give variety to the language used in your paper. Eliminate jargon and florid language; check to see that you have not used mixed metaphors; try to improve the quality of each sentence and each paragraph.

However, your main goal in rewriting should be to see that the flow of ideas from introduction to conclusion is smooth and unimpeded by extraneous material which interferes with the clarity of the exposition. You can make use of special devices to aid the reader to stay on the path along which you are leading him. Forecasting sentences and paragraphs will prepare the reader for later statements, while transitional statements connecting one idea to another will assist him in following your thought patterns. To make acceptable your final product, you must expend as much effort in revising as in preparing the original draft.

Illustrations

Photographs, maps, tables, and diagrams may be used in many different ways to illustrate and embellish a research report in geography. In fact, it is hard to conceive of a topic in this field that would not benefit from the use of illustrative material. However, the decision to use illustrations should not be a capricious one. The only justification for including any table, map, photograph, or diagram is to make your written argument clearer or more precise. Therefore, only use illustrations which further elucidate or enhance the discussion, and be sure to draw the reader's attention to each. Each illustration should be identified with a figure number and a title. Where needed, captions or legends explaining the illustration and tying it to the text should be added.

As a part of the original plan for your paper you should have some rough sketches of maps, tables, and diagrams. You also should give some thought to photographs that would give additional meaning to your words. As you proceed with your research, be on the lookout for illustrative materials that you could borrow or compile and when you are in the field taking notes, be prepared to use your camera to record examples of things that you wish to discuss in the text. As you are writing your paper, consider alternative methods of presenting information. Will a table be more useful than a graph or a map? In all of these considerations, remember that the preparation of illustrative materials is time-consuming; thus, the number of illustrations that can be prepared for a short paper is limited. In most cases, maps and diagrams must be relatively simple. See page 78 and pages 82–83 for further information on illustrations.

BORROWED MATERIALS

In your library search you may have come across useful illustrations. Your principal problem with such materials is to get a copy of them at the right scale for your report. If you are exceptionally lucky and find suitable illustrations at the right scale, you can make copies on a Xerox or similar copying device. From the student viewpoint this method of acquiring illustrative materials is most desirable because it is cheap and quick, and Xeroxed material may be easily modified. However, it does tend to smudge if handled too often, and most photographs and shaded and patterned areas do not reproduce well.

71

If you want to change the scale of the illustration, you will need some photographic means of duplication. For many purposes, photostatic copies are permissible and are most economical. However, there are some difficulties in appreciably changing scale of an illustration. Enlargements will give fuzzy edges to lines and lettering; reduction will sharpen lines but may make lettering too small to read. Photographs are usually printed with screen patterns consisting of many small dots. When enlarged greatly, the screened dot patterns become larger and farther apart, and may no longer carry the visual impression they did in the original.

In addition to the possibility of reproducing materials from books and atlases, you should also consider purchasing base maps and graph paper. Look in the yellow pages of your telephone directory under "Maps" to locate the store most likely to have the kind of map you want. In particular, the topographic maps printed by the U. S. Geological Survey make good base maps on which to place other data. Part II of this guide lists sources of maps of all kinds.

Remember that all borrowed materials belong to someone else. You need to get permission of the publisher in order to use them, and as a form of professional courtesy you should contact the individual who produced the original art work to ask him for permission also.

PHOTOGRAPHS

The value of photographs as a means of presenting geographical information should be apparent. In many instances, however, they are used poorly. Authors finish their reports, and simply tack on a few photographs because they feel it is expected of them. Photographs should be incorporated into the first general research plan and should serve a useful purpose. Photographs that you have taken in the field serve two purposes: they may record information which you may utilize in writing your paper, or they may amplify or clarify points that you make in your text. If you are writing a paper on the central business district of Denver, what better way is there to give your reader a "feel" for your study area than to show him some pictures of the area. However, there are a few words of caution in the use of photographs. They convey so much information that it is important to indicate precisely why they appear in your report. This may be done by writing extended captions or by commenting on them directly in the body of the report. For additional sources of photographs see pages 186ff.

COMPILATION OF MAPS,
TABLES, AND DIAGRAMS

In all likelihood you will be unable to locate illustrative materials that fulfill your needs and you will have to compile and draft your own maps, tables, and diagrams. An extensive discussion of the techniques of drafting and of graphic design is outside the scope of this guide. However, a few general suggestions may aid the beginning student in the preparation of simple illustrative materials for his research report. Should you wish additional guidance consult Greenhood's *Mapping* and Schmid's *Graphic Presentation*.[1]

The first and primary consideration in the preparation of an illustration is a clear-cut understanding of its purpose. For example, if it is to show the distribution of steel mills in the United States, show just that and do not add a half dozen additional bits of information. A good rule of thumb for beginning cartographers is to use single-purpose maps and diagrams and leave more complex cartographic problems for experienced workers in the field.

Once you have determined the purpose of your map or diagram, the next step is to acquire the necessary base map or graph paper and the data needed to construct the illustration. There is nothing unprofessional about copying or tracing material from a published source. Compilation involves the utilization of all available information on the topic on which you are preparing a map or diagram.

[1] David Greenhood, *Mapping* (Chicago: The University of Chicago Press, 1964). Calvin F. Schmid, *Handbook of Graphic Presentation* (New York: Ronald Press, 1954).

However, you do need to be certain that the material you are tracing is correct, and you must credit the source of your data.

One of the principal problems in using a base map is finding one of precisely the right size. You will probably have to enlarge or reduce the map to fit the dimensions of your paper. For most undergraduate reports, the best size for the final map is either 8½ x 11 inches or 11 x 16 inches so that it can be folded once. If you need to include a very large-scale map, you may fold it and place it in a pocket at the back of your report. Enlarging or reducing the scale of your map is not too difficult if your department has a Focalmatic Projector or a pantograph. If you are going to need a number of copies of the base map, make your tracing at the original scale and have copies made photographically at the desired scale.

If your finances are limited and a pantograph or Focalmatic Projector is not available, you may redraw the map to scale by using the method of equivalent squares (see figure). With a soft pencil construct a series of squares on tracing paper and lay it over your original map. Now lay a second piece of tracing paper over a grid with the same number of squares only in the size needed to produce a map of the desired scale. If you are reducing, the squares on the second grid will be smaller; if enlarging, they will be bigger. Then transfer square by square to the second grid all the necessary information on the original map. You are now ready to proceed with the drafting of your map.

Draw the frame which represents the working border of your map. Within it construct a legend box, a direction arrow, and, when needed, a locational grid consisting of meridians and parallels. In most cases it is necessary only to indicate the map grid along the borders. With a soft lead pencil or a blue drafting pencil enter all necessary information, including place names, onto your map.

If everything appears to be satisfactory, proceed to go over your map in ink. In most cases, lettering should be done before any of the line work or before you attempt to identify areas on your map. Parts of lines or areas are ordinarily not missed, but letters in a word would be; so, in most instances, lettering should have the

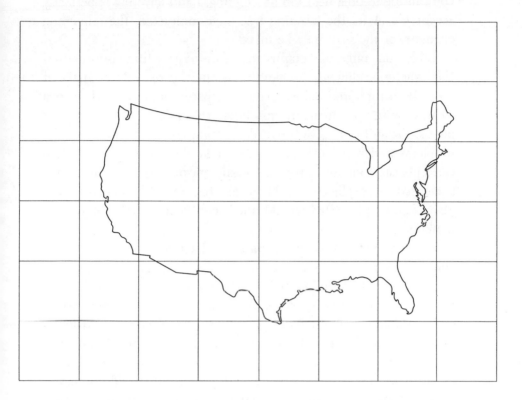

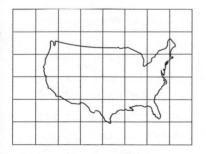

highest priority. Although there are a number of mechanical devices you can use to do a neat job of lettering, hand lettering is perfectly acceptable. After the lettering has been completed, the remaining elements of the map may be inked.

Tables, diagrams, and graphs are means of presenting information that might be difficult to handle within the text. They make it possible to interpret, analyze, and compare quantitative data in a clear, concise way. In many instances tables, diagrams, and graphs enable the author to give the reader a comprehensive picture of the problems and relationships developed in the report. In addition, concepts can sometimes be more easily grasped and remembered if presented in graphic form. There are few research reports in the field of geography which would not benefit from tabular or graphic material.

Choosing the right kind of graph or diagram sometimes is difficult. Before starting the actual work of designing and laying out an illustration look at the nature of the data and consider how it can best be shown. Then make several quick sketches of the various ways in which your material might be presented in a particular situation. Can your data be presented most usefully in tabular form or in graphic form? Your decision should be based on how well each form carries the information you wish to present. In addition, you need to consider the time required to prepare a graph or diagram. Never design or draft an illustration in a hurried or slipshod manner; rather it is better to include a smaller number of simple pieces, neatly done.

In constructing line and bar graphs use a piece of graph paper to construct a preliminary pencil version. Frame-in the area you will read and sketch in the title and the lettering for the horizontal and vertical scales. Plot your data on the grid, and then complete any lettering or line work in pencil. Now place a sheet of tracing paper or cloth over the preliminary graph and trace it in India ink. If several lines appear on your graph, vary the thickness or character of each.

The Final Draft

Many a good research effort has gone down the drain in a frantic last minute attempt to complete it before a deadline is reached. Therefore be sure to allow time for a thorough examination of your manuscript and illustrative materials to see that they contain no errors and are internally consistent.

You were advised earlier to find out your instructor's specifications before starting to work. Thus, the following section on contents should be used with discretion. Not everything mentioned needs to be included in all papers.

CONTENTS OF YOUR PAPER

Ordinarily, your paper will contain all or most of the following parts, arranged in the order given. Each should begin on a separate sheet of paper.

1. Title Page
2. Abstract
3. Preface or Acknowledgments
4. Table of Contents
5. List of Illustrative Material
6. Text
7. Footnotes
8. Appendices
9. Bibliography

1. The *title page* should contain the following: (a) title of the paper, in capital letters; (b) your name; (c) course name and number; (d) semester and year in which course is being taken; (e) date of submission of paper. Each piece of information should start a new line, which is centered on the page.

2. The *abstract* should inform the reader, in as few words as possible, about the important findings and conclusions of your paper. It should be written in the active voice and contain the key words found in your topic headings. It should make positive statements and set down factual conclusions. It should be typed single-space, on a shorter line than the text, and centered on the page immediately following the title page.

3. The *preface* should include the writer's statements of his reasons for making the study, the scope of the study, and the aid given him in making the study. However, if you wish to give recognition only to the agencies and individuals who have given you assistance, the heading *Acknowledgments* should be used.

4. The *table of contents* should contain all of the major topic headings in your paper. Whether or not you also include subtopics is a matter of choice which should be determined largely by the nature of your paper. All headings should appear precisely as they are in the body of your paper and should ordinarily be double spaced. However, if necessary, subtopic headings under the main headings may be single-spaced in order to place all of the table of contents on one page.

5. The *list of illustrative material* identifies by title all tables, photographs, maps, and charts and gives their location within the paper. In a short paper at the undergraduate level it is permissible to identify all of these as *figures* and number them in the order in which they occur. In longer papers, separate lists of tables, photographs, maps, and charts should be prepared and typed on separate sheets. The list of illustrative material should be similar in appearance to that of the table of contents, each piece carrying a number and a key phrase taken from its caption.

6. The *text* should be typed double-spaced except for quotations of three or more lines. These should be indented and single-spaced and should not be enclosed in quotation marks. Enclose all run-in quotations in quotation marks.

The title should be repeated on the first page of the text about two or three inches from the top of the page and separated from the body of the text by at least three spaces. Capitalize but do not underline the title and do not use a period after it.

Identify each footnote by typing the number associated with it a half-space above the line and after the complete statement being cited. Footnotes are numbered consecutively within an article and the numbers are placed after the punctuation marks.

Follow the instructions in the style sheet of the *Annals* (Appendix II) for topic headings. Be sure that you assign the proper heading values to equivalent segments of your paper.

Make use of a sample paper as a guide for typing purposes. Note the number of spaces for the indentation of the first sentence of a paragraph and the number of spaces between sentences. Remember to keep the margins on each page approximately the same. And be *neat.*

7. Type your *footnotes* on a separate sheet of paper and place them at the end of your paper. Double space each item cited and leave a double space between items. Check the section (pp. 65–66) in this chapter dealing with footnotes and Appendix II for the *Annals* style sheet.

8. The *appendix* is where you put material that is useful to the reader but that would disrupt the flow of your paper. Extensive tables, official documents, and letters are some of the materials that may be placed in the appendix. The appendix should come immediately after the footnotes and before the bibliography.

9. The *bibliography* will be an easy matter if you have followed the directions previously given on taking notes (pp. 39–40) because all of the references examined will be on 3″ x 5″ cards which can be alphabetized easily.

The bibliography should be single-spaced with a double space between items. Items should be alphabetized by author, with surname first. When no author is given, the first word of the title, excluding articles (*a, an, the*), is used to determine an item's location. Titles of books and journals should be underlined; titles of essays, articles within books and journals, theses, and dissertations should be enclosed in quotations.

If two or more titles by the same author are listed, a dash is usually substituted for the name of the author in each entry following the first one. See examples of form in the section (pp. 67–68) of this chapter on bibliography.

EDITING

Before typing the final draft of your paper, edit it carefully. Use one standard dictionary and style manual as a guide to correct spelling, grammar, and other matters of form. Many geographers use the style manual developed by the University of Chicago Press and suggest that students use the guides developed by one of its former editors, Kate L. Turabian. They are *A Manual for Writers of Term Papers, Theses, and Dissertations* and *Student's Guide for Writing College Papers.*[2] Final editing should be done in several different ways so that all possible errors are eliminated. It should be done section-by-section so that internal consistency can be achieved, and word-by-word, sentence-by-sentence, and paragraph-by-paragraph in great detail. It is a good idea to edit selectively for spelling, punctuation, and word usage. Check to see that all spelling and usage are correct and consistent. Go over sentence construction to see that verbs are in agreement with their subjects, that pronouns are in agreement with their antecedents, and that it is always clear what the antecedents are. Be sure that you do not have any incomplete sentences. In doing this final job of editing seize every opportunity to improve the language of your paper and to improve the transition from one thought to another. Finally, have some competent relative or friend look your manuscript over. Often, someone else with a detached eye will catch errors or unclear statements that may have escaped you.

[2] The University of Chicago Press, *A Manual of Style*, 12th ed. (Chicago: 1969); Kate L. Turabian, *A Manual for Writers of Term Papers, Theses, and Dissertations* (Chicago: The University of Chicago Press, 1967); and Kate L. Turabian, *Student's Guide for Writing College Papers* (Chicago: The University of Chicago Press, 1963).

TYPING

Whether the student does his own typing or has a typist do it, it is his responsibility to see that the necessary guidelines are followed. Most instructors will recommend a style manual for use by their students. Graduate schools issue specific instructions on the preparation of theses and dissertations. It is the student's responsibility to discover what specifications to follow in preparing his manuscript. A few basic rules to be followed in typing a research report are listed below.

1. All material in the manuscript should be typed on high-quality white paper, using only one side of a sheet. Before starting to type, test the paper to be sure that erasures and corrections can be made neatly and easily on it.

2. Make certain that your typewriter ribbon produces a solid black image. Most college professors have suffered eyestrain at one time or another trying to read barely legible words imprinted by a worn out typewriter ribbon.

3. In all cases type and retain a carbon copy of your paper. Some professors insist on keeping a copy of a student's paper, and some professors have been known to misplace student papers. Be prepared!

4. Margins on your paper should be consistent from one page to the next. It is generally recommended that the left-hand margin be on the order of an inch and a half and the top, bottom, and right-hand margins be on the order of an inch. This provides enough space at the left to enclose your paper in a hand-cover binder, which may be purchased at any stationery store.

5. Headings such as *Abstract, Preface,* and *Contents* should be placed at least two inches from the top of the page and centered. If the material on such a page is unusually short, both the heading and all other typed material should be placed slightly above the center of the page.

6. Pages of prefatory material in your paper should be numbered

with small Roman numerals (i, ii, iii, iv, etc.), counting but not numbering the title page. Each page of the text, the appendix, and the bibliography should be numbered with an Arabic numeral located in the same position on each page.

7. Typographical errors should be erased and then corrected with the typewriter. If a page looks too messy as a result of many erasures and corrections, it should be retyped.

MOUNTING AND FOLDING ILLUSTRATIVE MATERIAL

In preparing illustrative material for your final draft remember that all photographs, maps, graphs, etc. are going into an 8½ x 11 inch dimension and, therefore, plan accordingly. However, if you are going to submit the paper for publication, try to anticipate any changes in scale which may take place so that you do not have to redesign your art work. Photos and maps submitted for publication should be loose (not cemented to paper) and should be numbered. Make a separate caption sheet, with each caption identified by number. Listed below are some general rules to follow in preparing illustrative materials for research reports. (See Appendix II on preparing illustrations for the *Annals*.)

1. Margins should be comparable to those used in the body of your text. All descriptive matter must be located within this margin unless you have chosen to bleed the illustration off one or more sides of the page.

2. All illustrations should be identified with figure numbers in the order of appearance.

3. All illustrations should have legends or captions which clearly explain the content of the illustration. These may be placed on sheets of paper facing the illustrations.

4. The source of each illustration should be given below or after the legend or caption.

5. All illustrative material should, if at all possible, be mounted on paper the same size as the typed portion of your report. If you

are preparing multiple copies of your paper, it may be more economical to print all illustrative materials on sheets of photo paper of the correct size.

6. Dry mounting tissue, which may be obtained at any photographic supply house, is the best material that can be used for attaching one sheet of paper to another. It is applied with a warm iron. White glue and rubber cement, if used correctly, are also acceptable.

7. If illustrative materials are oversized, they must be folded. Where the folded materials is to be placed within the body of the text, more than one fold is to be discouraged. The reader can open and use an illustration more easily if it is placed in a pocket at the back of the report. Practice first with a sheet of paper the same size to discover the best way to fold your illustration.

SUMMARY

Writing a paper should be accomplished in three principal steps: planning, preparing a rough draft, and preparing the final manuscript. Planning should involve the time element, the inclusion of illustrative materials, and the preparation of a working outline. In preparing the rough draft, proceed section by section to fill in the framework created by your outline. Effective research and a systematic method of recording data will give you the raw material with which to construct your report. Skill in fashioning sentences and paragraphs and in carrying your reader from one thought to another in logical fashion will result in accomplishing the purpose of your paper.

Rewrite your paper as many times as is necessary to make the flow of ideas from the introduction to the conclusion follow a clear trail free of verbal obstructions. The list of illustrations should be completed early enough in the period of report-writing so that references to them may be included in the text. Maps and charts should be neat and simple; single-purpose illustrations are to be preferred over multiple-purpose ones. The final draft of your paper should follow some accepted model. If none is specified by your instructor, make use of an article from the *Annals*.

Problems for Study

1. For a term paper you are writing for another class prepare a working outline. Include a tentative list of tabular and illustrative materials.

2. From papers appearing in one of the major geographical journals select three articles that you feel are well organized and well written. Defend your choices.

3. Briefly define structure and style. In what way are they significant aspects of writing a report?

4. Examine the footnotes used in journals outside the field of geography. How do they compare with the footnotes that appear in the *Annals?* What are some of the principal considerations in preparing footnotes?

APPENDIX I

The Four Traditions of Geography*

by William D. Pattison

In 1905, one year after professional geography in this country achieved full social identity through the founding of the Association of American Geographers, William Morris Davis responded to a familiar suspicion that geography is simply an undisciplined "omnium-gatherum" by describing an approach that as he saw it imparts a "geographical quality" to some knowledge and accounts for the absence of the quality elsewhere.[1] Davis spoke as president of the AAG. He set an example that was followed by more than one president of that organization. An enduring official concern led the AAG to publish, in 1939 and in 1959, monographs exclusively devoted to a critical review of definitions and their implications.[2]

Every one of the well-known definitions of geography advanced

* Paper presented at the opening session of the annual convention of the National Council for Geographic Education, Columbus, Ohio, November 29, 1963. Reprinted with permission.

[1] William Morris Davis, "An Inductive Study of the Content of Geography," *Bulletin of the American Geographical Society*, Vol. 38, No. 1 (1906), 71.

[2] Richard Hartshorne, *The Nature of Geography*, Association of American Geographers (1939), and idem., *Perspective on the Nature of Geography*, Association of American Geographers (1959).

since the founding of the AAG has had its measure of success. Tending to displace one another by turns, each definition has said something true of geography.[3] But from the vantage point of 1964, one can see that each one has also failed. All of them adopted in one way or another a monistic view, a singleness of preference, certain to omit if not to alienate numerous professionals who were in good conscience continuing to participate creatively in the broad geographic enterprise.

The thesis of the present paper is that the work of American geographers, although not conforming to the restrictions implied by any one of these definitions, has exhibited a broad consistency, and that this essential unity has been attributable to a small number of distinct but affiliated traditions, operant as binders in the minds of members of the profession. These traditions are all of great age and have passed into American geography as parts of a general legacy of Western thought. They are shared today by geographers of other nations.

There are four traditions whose identification provides an alternative to the competing monistic definitions that have been the geographer's lot. The resulting pluralistic basis for judgment promises, by full accommodation of what geographers do and by plain-spoken representation thereof, to greatly expedite the task of maintaining an alliance between professional geography and pedagogical geography and at the same time to promote communication with laymen. The following discussion treats the traditions in this order: (1) a spatial tradition, (2) an area studies tradition, (3) a man-land tradition and (4) an earth science tradition.

Spatial Tradition

Entrenched in Western thought is a belief in the importance of spatial analysis, of the act of separating from the happenings of

[3] The essentials of several of these definitions appear in Barry N. Floyd, "Putting Geography in Its Place," *The Journal of Geography*, Vol. 62, No. 3 (March 1963), 117–120.

experience such aspects as distance, form, direction and position. It was not until the 17th century that philosophers concentrated attention on these aspects by asking whether or not they were properties of things-in-themselves. Later, when the 18th century writings of Immanuel Kant had become generally circulated, the notion of space as a category including all of these aspects came into widespread use. However, it is evident that particular spatial questions were the subject of highly organized answering attempts long before the time of any of these cogitations. To confirm this point, one need only be reminded of the compilation of elaborate records concerning the location of things in ancient Greece. These were records of sailing distances, of coastlines and of landmarks that grew until they formed the raw material for the great *Geographia* of Claudius Ptolemy in the 2nd century A.D.

A review of American professional geography from the time of its formal organization shows that the spatial tradition of thought had made a deep penetration from the very beginning. For Davis, for Henry Gannett and for most if not all of the 44 other men of the original AAG, the determination and display of spatial aspects of reality through mapping were of undoubted importance, whether contemporary definitions of geography happened to acknowledge this fact or not. One can go further and, by probing beneath the art of mapping, recognize in the behavior of geographers of that time an active interest in the true essentials of the spatial tradition —*geometry* and *movement*. One can trace a basic favoring of movement as a subject of study from the turn-of-the-century work of Emory R. Johnson, writing as professor of transportation at the University of Pennsylvania, through the highly influential theoretical and substantive work of Edward L. Ullman during the past 20 years and thence to an article by a younger geographer on railroad freight traffic in the U.S. and Canada in the *Annals* of the AAG for September 1963.[4]

One can trace a deep attachment to geometry, or positioning-and-layout, from articles on boundaries and population densities in

[4] William H. Wallace, "Freight Traffic Functions of Anglo-American Railroads," *Annals of the Association of American Geographers*, Vol. 53, No. 3 (September, 1963), 312–331.

early 20th century volumes of the *Bulletin of the American Geographical Society*, through a controversial pronouncement by Fred K. Schaefer in 1953 that granted geographical legitimacy only to studies of spatial patterns [5] and so onward to a recent *Annals* report on electronic scanning of cropland patterns in Pennsylvania.[6]

One might inquire, is discussion of the spatial tradition, after the manner of the remarks just made, likely to bring people within geography closer to an understanding of one another and people outside geography closer to an understanding of geographers? There seem to be at least two reasons for being hopeful. First, an appreciation of this tradition allows one to see a bond of fellowship uniting the elementary school teacher, who attempts the most rudimentary instruction in directions and mapping, with the contemporary research geographer, who dedicates himself to an exploration of central-place theory. One cannot only open the eyes of many teachers to the potentialities of their own instruction, through proper exposition of the spatial tradition, but one can also "hang a bell" on research quantifiers in geography, who are often thought to have wandered so far in their intellectual adventures as to have become lost from the rest. Looking outside geography, one may anticipate benefits from the readiness of countless persons to associate the name "geography" with maps. Latent within this readiness is a willingness to recognize as geography, too, what maps are about—and that is the geometry of and the movement of what is mapped.

Area Studies Tradition

The area studies tradition, like the spatial tradition, is quite strikingly represented in classical antiquity by a practitioner to

[5] Fred K. Schaefer, "Exceptionalism in Geography: A Methodological Examination," *Annals of the Association of American Geographers*, Vol. 43, No. 3 (September, 1953), 226–249.

[6] James P. Latham, "Methodology for an Instrumented Geographic Analysis," *Annals of the Association of American Geographers*, Vol. 53, No. 2 (June, 1963), 194–209.

whose surviving work we can point. He is Strabo, celebrated for his *Geography* which is a massive production addressed to the states-men of Augustan Rome and intended to sum up and regularize knowledge not of the location of places and associated carto-graphic facts, as in the somewhat later case of Ptolemy, but of the nature of places, their character and their differentiation. Strabo exhibits interesting attributes of the area studies tradition that can hardly be overemphasized. They are a pronounced tendency toward subscription primarily to literary standards, an almost omnivorous appetite for information and a self-conscious companionship with history.

It is an extreme good fortune to have in the ranks of modern American geography the scholar Richard Hartshorne, who has pondered the meaning of the area studies tradition with a legal acuteness that few persons would challenge. In his *Nature of Geography*, his 1939 monograph already cited,[7] he scrutinizes ex-haustively the implications of the "interesting attributes" identified in connection with Strabo, even though his concern is with quite other and much later authors, largely German. The major literary problem of unities or wholes he considers from every angle. The Gargantuan appetite for miscellaneous information he accepts and rationalizes. The companionship between area studies and history he clarifies by appraising the so-called idiographic content of both and by affirming the tie of both to what he and Sauer have called "naively given reality."

The area studies tradition (otherwise known as the chorographic tradition) tended to be excluded from early American professional geography. Today it is beset by certain champions of the spatial tradition who would have one believe that somehow the area studies way of organizing knowledge is only a subdepartment of spatialism. Still, area studies as a method of presentation lives and prospers in its own right. One can turn today for reassurance on this score to

[7] Hartshorne's 1959 monograph, *Perspective on the Nature of Geography*, was also cited earlier. In this later work, he responds to dissents from geographers whose preferred primary commitment lies outside the area studies tradition.

practically any issue of the *Geographical Review*, just as earlier readers could turn at the opening of the century to that magazine's forerunner.

What is gained by singling out this tradition? It helps toward restoring the faith of many teachers who, being accustomed to administering learning in the area studies style, have begun to wonder if by doing so they really were keeping in touch with professional geography. (Their doubts are owed all too much to the obscuring effect of technical words attributable to the very professionals who have been intent, ironically, upon protecting that tradition.) Among persons outside the classroom the geographer stands to gain greatly in intelligibility. The title "area studies" itself carries an understood message in the United States today wherever there is contact with the usages of the academic community. The purpose of characterizing a place, be it neighborhood or nation-state, is readily grasped. Furthermore, recognition of the right of a geographer to be unspecialized may be expected to be forthcoming from people generally, if application for such recognition is made on the merits of this tradition, explicitly.

Man-Land Tradition

That geographers are much given to exploring man-land questions is especially evident to anyone who examines geographic output, not only in this country but also abroad. O. H. K. Spate, taking an international view, has felt justified by his observations in nominating as the most significant ancient precursor of today's geography neither Ptolemy nor Strabo nor writers typified in their outlook by the geographies of either of these two men, but rather Hippocrates, Greek physician of the 5th century B.C. who left to posterity an extended essay, *On Airs, Waters and Places*.[8] In this work, made up

[8] O. H. K. Spate, "Quantity and Quality in Geography," *Annals of the Association of American Geographers*, Vol. 50, No. 4 (December, 1960), 379.

of reflections on human health and conditions of external nature, the questions asked are such as to confine thought almost altogether to presumed influence passing from the latter to the former, questions largely about the effects of winds, drinking water and seasonal changes upon man. Understandable though this uni-directional concern may have been for Hippocrates as medical commentator, and defensible as may be the attraction that this same approach held for students of the condition of man for many, many centuries thereafter, one can only regret that this narrowed version of the man-land tradition, combining all too easily with social Darwinism of the late 19th century, practically overpowered American professional geography in the first generation of its history.[9] The premises of this version governed scores of studies by American geographers in interpreting the rise and fall of nations, the strategy of battles and the construction of public improvements. Eventually this special bias, known as environmentalism, came to be confused with the whole of the man-land tradition in the minds of many people. One can see now, looking back to the years after the ascendancy of environmentalism, that although the spatial tradition was asserting itself with varying degrees of forwardness, and that although the area studies tradition was also making itself felt, perhaps the most interesting chapters in the story of American professional geography were being written by academicians who were reacting against environmentalism while deliberately remaining within the broad man-land tradition. The rise of culture historians during the last 30 years has meant the dropping of a curtain of culture between land and man, through which it is asserted all influence must pass. Furthermore work of both culture historians and other geographers has exhibited a reversal of the direction of the effects in Hippocrates, man appearing as an independent agent, and the land as a sufferer from action. This trend as presented in published research has reached a high point in the collection of papers titled *Man's*

---

[9] Evidence of this dominance may be found in Davis's 1905 declaration: "Any statement is of geographical quality if it contains . . . some relation between an element of inorganic control and one of organic response" (Davis, *loc. cit.*).

Role in Changing the Face of the Earth. Finally, books and articles can be called to mind that have addressed themselves to the most difficult task of all, a balanced tracing out of interaction between man and environment. Some chapters in the book mentioned above undertake just this. In fact the separateness of this approach is discerned only with difficulty in many places; however, its significance as a general research design that rises above environmentalism, while refusing to abandon the man-land tradition, cannot be mistaken.

The NCGE seems to have associated itself with the man-land tradition, from the time of founding to the present day, more than with any other tradition, although all four of the traditions are amply represented in its official magazine, *The Journal of Geography*, and in the proceedings of its annual meetings. This apparent preference on the part of the NCGE members *for defining geography in terms of the man-land tradition* is strong evidence of the appeal that man-land ideas, separately stated, have for persons whose main job is teaching. It should be noted, too, that this inclination reflects a proven acceptance by the general public of learning that centers on resource use and conservation.

Earth Science Tradition

The earth science tradition, embracing study of the earth, the waters of the earth, the atmosphere surrounding the earth and the association between earth and sun, confronts one with a paradox. On the one hand one is assured by professional geographers that their participation in this tradition has declined precipitously in the course of the past few decades, while on the other one knows that college departments of geography across the nation rely substantially, for justification of their role in general education, upon curricular content springing directly from this tradition. From all the reasons that combine to account for this state of affairs, one may, by selecting only two, go far toward achieving an understanding of this tradition. First, there is the fact that American college geography, growing out of departments of geology in many crucial instances, was at one time greatly overweighted in favor of earth science, thus rendering the field unusually liable to a sense of loss as better balance came into being. (This one-time disproportion found reciprocated support for many years in the narrowed, environmentalistic interpretation of the man-land tradition.) Second, here alone in earth science does one encounter subject matter in the normal sense of the term as one reviews geographic traditions. The spatial tradition abstracts certain aspects of reality; area studies is distinguished by a point of view; the man-land tradition dwells upon relationships; but earth science is identifiable through concrete objects. Historians, sociologists and other academicians tend not only to accept but also to ask for help from this part of geography. They readily appreciate earth science as something physically associated with their subjects of study, yet generally beyond their competence to treat. From this appreciation comes strength for geography-as-earth-science in the curriculum.

Only by granting full stature to the earth science tradition can one make sense out of the oft-repeated addage, "Geography is the mother of sciences." This is the tradition that emerged in ancient Greece, most clearly in the work of Aristotle, as a wide-ranging study of natural processes in and near the surface of the earth. This is the tradition that was rejuvenated by Varenius in the 17th century as "Geographia Generalis." This is the tradition that has been subjected to subdivision as the development of science has approached the present day, yielding mineralogy, paleontology, glaciology, meteorology and other specialized fields of learning.

Readers who are acquainted with American junior high schools may want to make a challenge at this point, being aware that a current revival of earth sciences is being sponsored in those schools by the field of geology. Belatedly, geography has joined in support of this revival.[10] It may be said that in this connection and in others, American professional geography may have faltered in its adherence to the earth science tradition but not given it up.

In describing geography, there would appear to be some advantages attached to isolating this final tradition. Separation improves the geographer's chances of successfully explaining to educators why geography has extreme difficulty in accommodating itself to social studies programs. Again, separate attention allows one to make understanding contact with members of the American public for whom surrounding nature is known as the geographic environment. And finally, specific reference to the geographer's earth science tradition brings into the open the basis of what is, almost without a doubt, morally the most significant concept in the entire geographic heritage, that of the earth as a unity, the single common habitat of man.

An Overview

The four traditions though distinct in logic are joined in action. One can say of geography that it pursues concurrently all four of them. Taking the traditions in varying combinations, the geographer can explain the conventional divisions of the field. Human or cultural geography turns out to consist of the first three traditions applied to human societies; physical geography, it becomes evident, is the fourth tradition prosecuted under constraints from the first and second traditions. Going further, one can uncover the meanings of "systematic geography," "regional geography," "urban geography," "industrial geography," etc.

It is to be hoped that through a widened willingness to conceive of and discuss the field in terms of these traditions, geography will be better able to secure the inner unity and outer intelligibility to

[10] Geography is represented on both the Steering Committee and Advisory Board of the Earth Science Curriculum Project, potentially the most influential organization acting on behalf of earth science in the schools.

which reference was made at the opening of this paper, and that thereby the effectiveness of geography's contribution to American education and to the general American welfare will be appreciably increased.

APPENDIX II

Editorial Policy Statement and Style Sheet of the Annals, Association of American Geographers

Editorial Policy Statement[1]

General Commentary

For the benefit of members of the Association, an incoming Editor of the *Annals* should explain the changes which he intends to make, the policies by which he will be guided, and the procedures which he intends to follow. Our journal, in attaining the ripe age of three-

SOURCE: Reprinted from *Annals*, Association of American Geographers, Vol. 60, No. 1 (March, 1970), pp. 194–207, with permission.

[1] A free copy of the Editorial Policy Statement, including the Style Sheet, may be had by writing the Editor.

96

score years, has established a format and a pattern of content which most of us find satisfactory, and change should not be made solely for the sake of change. Three changes will be made, however, in the belief that they will enhance the interest and value of the journal for its readers: 1) The Review Section, under the Editorship of Edwin H. Hammond, will be expanded to include reviews of individual books, and will continue to have Review Articles; 2) for the benefit of the international community of scholars, most of whose members are accustomed to a system of measurement which has a somewhat more rational basis than the parts of the human body in medieval England, authors will be required to include metric equivalents in parentheses after all nonmetric measurements given in the texts of their manuscripts; and 3) shorter articles will be encouraged, both by the abolition of any restrictions, other than high scholarly quality, on the minimum length of articles, and by the imposition of a maximum length of twenty-five (25) *Annals* pages per article. Authors will be encouraged to present their ideas in lucid, succinct, well-written, readable prose.

Historically, the editors of this journal have sought to publish mature, substantive, scholarly studies which are representative of the breadth and depth of geography. There has been, and shall continue to be, a thoroughly catholic policy with regard to subject matter, with no bias either in favor of or against any particular topic. An editor cannot create scholarly works for his journal, however; he can only receive and process contributions submitted by productive scholars. Any lack of balance which may be perceived in the content of the *Annals* should and will reflect an imbalance in geographic scholarship, as indicated by the quality and quantity of material submitted, and not an editorial bias.

The primary duty of the Editor of the *Annals* is to maintain, and to aspire to enhance, a high level of scholarly quality in the organization and presentation of material, whatever the subject matter may be. The Editor will be happy to work with authors in developing contributions which will maintain the quality which has come

97

to be expected of articles which are published in this journal. The advice and assistance of qualified referees will be sought as and when necessary, but the Editor has the final responsibility for making the decision as to which materials shall be published.

The Review Section

It is anticipated that the Review Section, which will be expanded to include reviews of individual books, as well as Review Articles, will eventually occupy a major segment of each issue of the *Annals*. Correspondence concerning the Review Section, including suggestions, recommendations, and questions concerning reviews of individual books or review articles, and publications for review, should be addressed to EDWIN H. HAMMOND, *Annals* Review Editor, Department of Geography, Syracuse University, Syracuse, New York, 13210.

Map Supplements

The highly successful series of Map Supplements, which are inserted loose leaf in the back of the *Annals*, will be continued. Anyone who is considering the preparation of material for publication in this series should contact the Map Supplement Editor at the earliest possible stage in his planning. Information and advice concerning the size, text, layout, projections, preparation of copy, or any other aspect of the Map Supplement program may be obtained from NORMAN J. W. THROWER, *Annals* Map Supplement Editor, Department of Geography, University of California, Los Angeles, California, 90024.

Articles

No article will be published which requires more than twenty-five pages in the *Annals* once the present backlog of longer papers has been published. An *Annals* page carries roughly 800 words, and the length of the text must be adjusted to leave space for the number of illustrations which accompany it. No policy, other than the requirement of high scholarly quality, has been established concerning the minimum length of articles which will be accepted.

An article which is published in the *Annals* should speak to all geographers, not just to an esoteric in-group. It should clearly explain its objectives, its usefulness, and its relevance to the field of geography. Its content should be placed in context with related studies, and oriented toward a general understanding of its subject. Its thesis should be carefully documented by presentation of the full range of evidence pertinent to the problem dealt with. All maps, photographs, tables, diagrams, and documentary citations should be complete and of high quality. Articles which fail to meet the requirements of quality, in all respects, will be returned to their authors with suggestions for the improvements which are needed to make their publication possible.

THE ABSTRACT

Each article must be preceded by an Abstract, and any paper submitted without one will be returned to the author for completion. A descriptive abstract, which is written in the passive voice and merely describes the procedures followed in the paper, will not be accepted.[2] The Abstract must contain substantive information about the content of the paper. It should include the important

[2] A horrible example of a descriptive abstract: "The general hypothesis for the solution of this problem is presented through two alternative models. The evidence of a variety of tests of each model is presented, and the hypothesis is shown to be valid."

points which are made in the body of the text, employing the key words pertaining to the subject matter, and should present the gist of the conclusions. Such an abstract should not be confused with a summary, which is longer and more detailed, nor with a conclusion, which states the findings of the paper.

No precise statement can be made about the length of the Abstract, because it will vary with the nature and length of the paper, but 150 words is the maximum length permissible, and at least sixty words will be required for most papers. The format, placement, and typical phraseology of an Abstract are specified in the Style Sheet.

MAPS AND DIAGRAMS

The inclusion of documentation in the form of maps, diagrams, sketches, drawings, graphs, scattergrams, and other forms of graphic presentation normally is vital to a paper in geography, but when published such documentation often has suffered from inadequate preparation, both in type of illustration and in quality production of the original. The particular attention of intending authors is called to the Recommendations For the Preparation of Maps for Publication in the *Annals*, which was prepared by the Association of American Geographers Committee on Cartography (1964), to the section on The Preparation and Submission of Manuscripts, and to the Style Sheet.

An author whose paper is accepted for publication will be expected to submit original line drawings of all diagrammatic materials, because photostatic or photographic copies rarely permit quality reproduction, since blackness at the outer margins is not uniform. This is particularly true when several varieties of adhesive patterns, such as Zipatone, are used, since matching patterns of blackness can seldom be assured. What may look relatively even to the naked eye may become quite uneven in the reproduced painting cut.

An author should avoid using too many varieties of pattern on a single map, because it is difficult to produce legible results of high quality from complex maps which contain many patterns.

PHOTOGRAPHS

The inclusion of photographs, as a form of documentation of textual statements, is also normal in a geographic paper. Authors should be warned, however, against submitting too many photographs, photographs which are not essential, photographs whose quality is not commensurate with the quality of the text of the paper, or photographs which are not effectively captioned.

IDENTIFICATION OF FIGURES

Each illustration, including photographic illustrations, should be marked with a figure number and a key phrase taken from the title of the article. The *Annals* uses only one series identification for all forms of illustration; all illustrations are termed Figures. They must be numbered in the order in which they are to appear in the text, and each figure must be given a keying reference in the text at whatever point it is desired that it appear. Captions should not be attached to individual figures, but should be typed on a separate sheet or sheets which bear the title of the article only. The name of the author should not appear either on figure identifications or on caption lists.

Annals Commentary

The *Annals* invites, and will continue to publish, significant critical comments on articles whch have been published in this journal. Although they need not have an Abstract, comments intended for the *Annals* Commentary section should be formally prepared and submitted in the same form as any other article; informal letters to the Editor will not be considered appropriate to this section. The comments must have a title, and the opening sentence must identify the paper being discussed. Documentation must be supplied for all references to other literature, with footnotes fully cited and properly typed on sheets separate from the text. The name and institutional affiliation of the author should be placed at the end of his communication, on the right hand margin.

A critic should believe that his comments are sufficiently significant to merit publication, and he should be certain that he wishes to have them printed as submitted. A copy of his comments will be sent to the author of the article which he discusses, and the author will be invited to prepare a reply if he sees fit to do so. The original author's reply will not be sent to the critic (who might be tempted to shift his ground after reading it), but both comments and reply will be published together, and as promptly as possible.

Preparation and Submission of Manuscripts

Serious, responsible scholarship includes the careful preparation of a manuscript tailored to the format of the journal in which publication is sought. The author who submits a manuscript which is awkward in format, layout, and arrangement has shirked his responsibility, and attempting to shift to someone else the burden of making his material suitable for the printer. The Editor cannot be expected to do the author's work for him, and an improperly prepared manuscript will be returned to the author with the request that it be submitted in proper form.

As a general rule, the Editor will be guided, but not governed, by the rules in A *Manual of Style*, 12th revised edition (Chicago: University of Chicago Press, 1969). The following sections depart from that manual in various particulars.

PREPARATION OF THE MANUSCRIPT

Manuscripts must be typed *double spaced in every respect* on one side of sheets of 8½″ x 11″ (or British quarto) paper, with a left hand margin of one and one-half inches. Legal-length or foolscap-length paper should not be used. Either elite or pica type is permissible. The name of the author should *not* appear on any page of the manuscript, the footnotes, or the captions for the illustrative material, so that the paper may be judged on its quality alone if it is sent to a referee. Material complementary to the text of the manuscript should be identified by a key phrase taken from the title of the article.

The text of the manuscript should run continuously. All footnotes should be in a single numbered series, which should include all initial notes, comments, and acknowledgments. All tables, and all quoted material entered into the text or the footnotes, should be *double spaced* in the same manner as the text, with appropriate marks and sources noted, as indicated in the Style Sheet. All tables should be typed on sheets separate from the text, one table to a sheet or sheets, and numbered separately from the text page numbers. The footnotes themselves should be typed on separate sheets as a continuous series, numbered, and *double spaced* throughout, because the text and the footnotes are set in type separately and joined only when pages are made up. Captions for illustrative matter should also be typed on separate pages numbered in a continuous series, and should be *double spaced* throughout; captions should not be attached individually to the separate pieces of illustrative matter. Each piece of illustrative matter should be identified by a number and the key phrase taken from the title. The abstract should be placed as the first paragraph of the paper, immediately following the title, and should be inset an additional five spaces.

Each measurement that is stated in nonmetric units must be followed immediately in the text by its equivalent in units of the metric system, carried to the first decimal place, stated in Arabic numerals, and enclosed by parentheses. See the example in the Style Sheet.

SUBMISSION OF THE MANUSCRIPT

The author needs to submit only one typed copy of textual and complementary material, prepared as described in the preceding section. He should include copies of all graphic material at printing scale, as described in the last paragraph of the Cartography Committee's statement (Submitting Illustrations to the Editor). Preview copies (which may be negative-sized contact prints) of all photographs intended to accompany the article should also be included. The manuscript should be accompanied by a letter stating the affiliation and mailing address of the author, and containing any special information necessary to editorial judgment as to the readiness of the article for publication.

This letter must clearly certify that the paper has not been published previously in any other journal or book in literal or approximate form, and that it has not been and will not be submitted elsewhere until a decision is received from the Annals. *The* Annals *does not publish secondary materials, and a duplicated manuscript will not be considered for publication.*

Upon acceptance of the article the Editor will request the author to forward all necessary originals of illustrative material supporting the article, including original drawings of graphic materials and black-and-white glossy prints of photographs on the order of five by seven inch dimensions. When possible all maps should be mailed flat, since rolling tends to loosen stickup.

RECOMMENDATIONS FOR THE PREPARATION OF MAPS FOR PUBLICATION IN THE ANNALS

The following recommendations for the preparation of maps have been prepared by the Cartography Committee to serve as a guide for authors. They provide the *minimum standard* for editorial purposes.

Dimensions

Illustrations must be planned so that their dimensions are proportional to the *Annals* page. The present full-page format allows a maximum printed image of 5.8 inches by ca. 8.3 inches. Space within this area must be allowed for a figure number and caption. Smaller illustrations should either a) extend across the page the width of two columns, i.e., 5.8 inches, or b) occupy the width of a single column, i.e., 2.8 inches. An illustration that occupies only part of a page, whether of one- or two-column width, ordinarily should not have a depth, including caption, of more than ca. 6 inches so that several lines of text may be set below it. Each figure should be provided with a firm border indicating reproduction limits no greater than those stated above. In special cases, where maps are large enough and important enough to require a double-page spread, the maximum image area has a vertical dimension of ca. 8.3 inches and a horizontal dimension of 11.6 inches. However, the latter may be extended to ca. 12.9 inches if one wishes to exceed the inner margin in order to reduce the gap to the minimum required for binding.

These dimensions are dictated by the *Annals* format. They are basic requirements in the earliest stages of illustration preparation.

Layout

Compilation and drafting of illustrations are usually done at a scale greater than the publication scale. However, authors should be *extremely careful* if they intend to submit illustrations more than twice the scale of the printed map because 1) the problems of adequate design are greatly compounded when the fair drawing is much larger than twice reproduction scale; 2) finer elements of linework, patterns, and lettering or type frequently fail to photograph uniformly; and 3) large drawings multiply handling difficulties and transportation charges. Consequently, whenever possible compilation dimensions ought not to exceed twice those given above. To achieve consistent results in a given paper, it is extremely helpful to make all drawings for the same reduction.

Minimum Sizes of Lettering or Type

Lettering so small as to be essentially illegible is a very common error in illustrations. Minimum sizes of lettering which will reproduce satisfactorily are given in the table below. These are almost threshold requirements both from the point of view of the method of reproduction and readability and it is wiser to use lettering well above this minimum size. It is recommended that the heights of all lettering exceed $\frac{1}{20}$ of an inch at reproduced size and that lettering with hairlines or extremely fine serifs not be used.

MINIMUM SIZE OF LETTERING

	Printed size	If compilation size is 1.5 $\times$ printed size	If compilation size is 2 $\times$ printed size
Type	4 pt	6 pt	8 pt
LeRoy	40	60/000 pen	80/000 pen
Wrico	60–45	90/7T pen	90/7T pen

Shading Patterns

The differentiation of areas on black-and-white maps requires, in the majority of cases, the use of patterns of dots or lines. The most commonly used patterns are preprinted varieties readily available in the United States under the trade names Artist Aid, Artype, Craftint, Transograph, and Zipatone. Gross shading patterns, such as parallel lines or large dots, which are bothersome to the eye or which attract undue attention to themselves, should not be used. On the other hand, finer patterns must be carefully selected to insure that they are coarse enough to be retained through the photographic and printing processes. It should be kept in mind that patterns made up of fine lines or dots, or having fine interspaces, are difficult to reproduce and should be avoided.

Reproducibility

The *Annals* is printed by letterpress and this method of production imposes some limitations upon the illustrator. The process of preparing the letterpress plates requires the use of high-contrast film which is incapable of reproducing gray tones. Thus, the original or fair drawing, including all preprinted symbols, lettering, and patterns affixed to the drawing, must have clean, jet-black lines to insure good reproduction. Weakly printed base maps, poor diazo and photocopy prints, and inked lettering or linework that is gray will not reproduce satisfactorily.

In cases where the reproduction of a map requires the use of halftone screens, or successive exposures of the press plate, the editor should be consulted in advance since they may entail additional costs which will have to be absorbed by the author.

Map Components

It is clearly desirable that most maps should include a brief title, scale, frame of reference, and documentation of special information.

In special circumstances one or more of these items may be omitted but the author should omit them only when he is certain the map reader will not become confused by their absence. The titles should be brief and focus attention on the subject matter of the map. To provide orientation a geographic grid is usually preferable on small-scale maps, and a north arrow can be used on very large-scale maps. If distance or area understanding is in any way essential to the purpose of the map, a graphic scale should be included. If the natural scale is used, it must be calculated for the reproduced size. Neat-lines (firm borders) should be placed around all cartographic and diagrammatic materials submitted. Authors should be extremely careful that any stickup added to the face of a map is securely fastened down. It frequently happens that stickup is loose on maps when received by the editor, and that more comes off by the time the originals reach the printer. Problems can be prevented by careful work at time of preparation. Maps on acetate sheets should be interleaved by sheets of paper to prevent the buildup of static electricity, which may move ink and loosen stickup.

Submitting Illustrations to the Editor

Copies made from final or near-final draft versions of illustrations should accompany the manuscript when it is first submitted to the editor. Do *not* send original drawings at this time, but instead furnish black-and-white photographic copies at *printing scale* (or ozalid, photostat, or some other form of reproduction). The submittal of such printing scale copies is necessary for the determination of reproduction quality; it also permits the submittal of illustrations to a referee, when one is used, for the more effective judgment of the whole quality of the article. Each illustration should be marked with a figure number and a key phrase taken from the title of the article, but should not bear the name of the author. Figure captions should not be individually attached to each figure, but should be assembled on one sheet (or sheets if numerous figures are used) bearing the title of the article only. Author names should not appear either on figure identifications or on caption lists.

ASSOCIATION OF AMERICAN GEOGRAPHERS,
CARTOGRAPHY COMMITTEE,
RICHARD E. DAHLBERG, CHAIRMAN.

Style Sheet

Format of the Typescript for All Articles
(On 8½ x 11 inch, or British quarto, paper only)

Title.

GEOGRAPHY AND GEOGRAPHERS[1]

(Avoid long titles. Note-takers, indexers,
bibliographers, and students become weary struggling
with long titles. The shorter the title, normally,
the better the title; titles over twelve words will
normally be shortened by the editor. The title
should be informative concerning content in order to
provide bibliographic indexers and abstracters with
needed key words and accurate information about the
article.)

Abstract.
(Inset five
spaces deeper
on left hand
margin than
main text.)

ABSTRACT. The loss and abandonment
of agricultural land in thirty
-one states of the eastern United States has
been more widespread than is commonly recognized. Strip mining and the loss of a locally
dominant crop have been important factors in
certain areas. The Soil Bank program has had
its greatest impact upon land of intermediate
quality. Land acquisition by forest industry
companies has borne little relationship to the
loss and abandonment of farm land. In the East
as a whole it appears that physical hindrances
to effective agriculture have been the most
important factor influencing the loss and
abandonment of farm land. (This is not the
full abstract, but note that each sentence
makes a positive statement; no sentence says
that tests are devised, data are manipulated,
and conclusions are drawn.)

Start of Text.
(Left hand
margin of
1.5 inches.)

Geography is a subject discipline which has been
practiced by many kinds of observers, such as
explorers, professional scholars, students, and
interested amateurs for almost as long as man has
inhabited the earth.[2]

First-order
Heading.

THE GEOGRAPHICAL OUTLOOK

The single characteristic common to all geographers is their concept

Second-order
Heading.

Primary Geographical Concepts

There are a number of concepts held to be basic
and primary among

The Regional Concept

Third-order Heading.

The Regional Concept

Just as geography is concerned with the whole earth, it also is concerned with regional entities of various sizes.[3] These areal units

Fourth-order Heading.

The Theory of the Subregion.

In the discussion of regional patterns Jones clearly stated the case for subregions.[4]

Figure and Table references to be placed only at the ends of substantive sentences.

The maps illustrating subregions were based on duplicate field surveys by two different field workers (Fig. 1, Table 1). DO NOT WRITE: Figure 1 gives the results of field surveys, and the comparison is given in Table 1. ALL REFERENCE to illustrative material should be placed at the ends of substantive sentences, in the interests of economy of space.

Format for a Table. Each table must give data source. Tables are to be typed completely double spaced on sheets separate from text, and numbered separately, one table to a sheet, or sheets.

TABLE 1.--THE RELATIONSHIPS OF REGIONAL DATA

Area	Relief	Climate	Population Per Sq. Mi.	Crops
Coast	Low	Semiarid	50	None
Hills	Rolling	Humid	30	Mixed

Source: Author field survey, supplemented by data from Jones, op. cit., footnote 12.

Footnotes should be placed at the end of sentences for short quotations. Any quotations of four or more lines long in the typescript should be inset three spaces, typed double spaced, have the footnote precede the quotation, and quote marks should be omitted.

Brown suggested that much of the coastal population ". . . existed on a beachcombers' paradise . . ." and that daily working hours were few.[5] However, he went on to say that:

Much of the population of the coastal fringe seems to enjoy the rather interesting activity of wandering the beaches and tidal flats during the whole of the daylight hours, combining their beachcombing with sunbathing, frequent short naps, swimming, and bartering the haul among friends, visitors from inland localities, and occasional tourists.

Do not give each source a separate number. Read the accompanying instruction for handling several sources with one footnote, and see footnote 7 in the section on footnotes.

When it is necessary to quote several authors for the statements made in a single sentence, such as Long, Brown, Slimm, and White, the footnote should still come at the end of the sentence.[7] In the footnote itself, the authors may be cited in the order given above, and explanatory phrases may be used to distinguish the aspect dealt with by each author. Often, the titles themselves make this clear.

Form and placement for listing items.	may be listed as a combination of causes such as the following: 1) Insufficient data bearing on specific factors; 2) divided systems of local controls; 3) continued migration and re-migration.

Miscellaneous Style Notations Applicable
to Textual Material

Precise date.	The first occurrence took place on August 17, 1904, during the . . .;
General dates.	Between 1951 and 1960; during the years, 1952-1954; in the early 1920's and again in the thirties; during the nineteenth century. Do not use 1919-27 or 19th century as forms.
Small numbers.	Numbers involving two digits only should be spelled out, as fourteen and twenty-six, except in technical discussions involving their frequent usage, as shown below in handling percentages.
Large numbers.	Numbers involving three or more digits should be given in Arabic numerals, as 150, 300, and 750; for four digits or over the full notation is needed, as 2,131, and 15,050,200.
Percent and Percentages.	The fraction amounted to thirty percent for the year; during the months of April, May, June, and July the percentages were 16, 3, 12, and 14, respectively. The word percent is an adverb and not a noun. Say "the percentage of the population," not "the percent of the population."
Punctuating a series.	The boy ate salad, meat, potatoes, asparagus, and ice cream for supper, but he did not eat salad, meat, potatoes, asparagus and ice cream for supper since he did not like asparagus in his ice cream. The sequence is 1, 2, 3, 4, and 5; not 1, 2, 3, 4 and 5.
Ellipsis.	In the conclusion . . . a statement of basic principles showed that; " . . . the third reason is the significant one;" "When . . . structural controls are clearly orogenetic . . . " the case becomes as King stated it; as Wallace put it: "There is no reason to suspect that the population remained stable during the whole period . . . ." (In the last example the fourth dot is a period ending the sentence.)
Use of the dash.	A dash is typed as two hyphens--when it must be used--without spacing, but a better phrasing is: when it must be used a dash is typed as two hyphens without spacing; avoid dashes in textual material by finding a better form of construction.

111

Dieresis, and other symbols.	Insert all proper diacritical marks as used in the particular language being employed. Köppen is preferable to Koeppen in most German surnames and place-names; Göring was normal, as is Göttingen, but Goebbels was normal in English.
Use of italics.	The second, third, and fourth orders of textual heading should be italicized, and italics are properly used in bibliographical entries, but do not use italics for emphasis in textual writing. All foreign words should be italicized.
Foreign words.	The term jhum is an Indian term for shifting cultivation and the Philippine term is caingin; American dry farming has similar practices. All foreign words should be italicized, but use a dictionary to be certain that they are not English forms.
Abbreviations.	Abbreviations should not be used, in order that clarity of meaning is evident to all readers, in whatever country they reside. Particularly, political territorial terms should not be abbreviated, since the local regional system may not be clearly evident to readers in another country. In straight textual reference the word Figure is written out, whereas in citing an illustration the abbreviation is used (Fig. 1). The word Table should always be written out. No abbreviations should be used in footnotes, since full citation of the bibliographic reference is required, except that authors are to be cited by initials only--only do not use full names except when full names are needed to distinguish two authors having the same name and initials.
Mathematical symbolization.	Authors who plan to use mathematical symbols in their manuscripts should carefully study Chapter 13, with particular attention to pages 306 and 307, of A Manual of Style, 12th revised edition (Chicago: University of Chicago Press, 1969).
Metrification.	Metric equivalents must be inserted in parentheses after all nonmetric measurements given in the text, thus: A healthy 150 pound (68.1 kg.) male should be able to run four miles (6.4 km.) in less than an hour.
Words to watch in usage.	The word "data" is plural; datum is singular. Data are, NOT data is. The word "while," despite much current misuse, is a conjunction meaning "during the time that." It is not a proper substitute for "whereas," "although," "but," "as," or "and." And "While at the same time that . . ." is a sheer redundancy. The word "whilst" is obsolete in the United States. The phrase "due to," in ascribing cause is still poor usage, despite its common application. The word "due" is an adjective carrying

the sense of obligation, be it moral, filial, financial, or legal. It also is proper in such terms as "due west," by long historic usage. Various forms of "owe" as "owing to," "owes," or "owed," are preferable, but there are other words denoting cause that may be used in place of "due to."

Use of present and past tenses.

In dealing accurately with geographic conditions the present tense should not be used to describe events which are long past. This applies not only to geographic conditions but also to the beliefs and statements attributed to authors who published years ago. Any reference to an author's statements, beliefs, or position, if more than three years old, must be put in the past tense. It is correct to say: "The book by Jones contains the phrase: . . .," because the book may contain that phrase, but it is not correct to say: "Jones says . . .," "Jones states . . .," or "Jones believes . . .," when Jones wrote his comments years ago and has since altered his opinion.

Spelling rules.

The Annals is an American journal, and it will retain "American English" spelling. Authors outside the United States and Canada will save themselves trouble if they will initially spell in the American manner. "American English" has abandoned the "u" in such words as harbor, labor, neighborhood, and favor; has transposed the "re" to "er," so that centres are centers, metres are meters, calibre is caliber, and spectre is specter; has used the "z" in place of the "s," so that it is civilization, utilize, analyze(but analysis), hypothesize, criticize, commercialize, anglicize, and so on.

The use of the hyphen.

Authors will save themselves trouble if they will consult their dictionaries, and exercise care in the use of the hyphen. Most of the "non" words, such as nonalcoholic, nonexclusive, nonracial, nonelective, nonofficial, nontaxable, nonterritorial, nonmetric, and nonworker are written without the hyphen. A great many of the words employing the prefixes "inter" and "intra" are also written without the hyphen. The same applies to the prefixes "extra," and "sub." A hyphen placed at the end of a line of typescript means that the end word is to be joined to the first word in the next line WITHOUT A HYPHEN. Authors should be careful not to end a line with a term such as one-third, in which the "one" is the last word in one line and "third" the first word in the succeeding line, since this tells the typographer to join the two words without a hyphen. In such cases, the hyphen is properly placed in the first space in the succeeding line (see the first and second lines of the sample abstract above).

Fractions and written-out numbers, such as one-half, twenty-three, and sixty-nine should always be hyphenated.

The use of quotation marks.

Words that require quotation marks within sentences must have double marks, thus: Loosely called "nationalism," the chief "reasoned deductions" of such experts. Short quotations of less than four textual lines must be enclosed by quotation marks; quotations of four or more textual lines should be inset three spaces and have quotation marks omitted. A word or phrase inside a surrounding pair of quotation marks takes only a single quotation mark before and after the word or phrase. Titles of articles from journals should be surrounded by quotation marks; separately published items are to have the titles underlined, to indicate italics. Punctuation

Punctuation related to quotation marks.

marks should be placed inside quotation marks, thus: B. Jones, "Urban Structure;" loosely called "nationalism," and " . . . the end of an era."[4] This includes such cases as " . . . than in the economic field?"

Style Notations Applicable to Bibliographic Footnotes and to Figure Captions

Note that in the typescript copy no period is placed immediately after the number, either in the text number or in the tabulation of the footnotes, and that the footnote number is set out beyond the text margin. Footnotes must be typed double spaced on a sheet or sheets separate from the text.

FOOTNOTES--GEOGRAPHY AND GEOGRAPHERS

An article published in a larger publication should have its title in quotation marks, and the title of the larger publication should be underlined, to indicate italic type. Separate publications, including separates from state and federal offices, should have their titles underlined, followed by full details of publication in parentheses.

Initial note, if needed.

1 Acknowledgment is given to the University of Greenland for support in carrying out the following study during a sabbatical leave.

2 No precise definition is here given to the age of man on the earth.

3 For a theoretical discussion of the bases of regionalism, see B. A. Jones, The Case for Regionalism (New York: Oxbridge Bookhouse, 1960), pp. 21-29.

Article citation.

4 U. R. Jones, "The Region and the Sub-Region," Records of Geography, Vol. 1 (1950), pp. 28-58, 137-57, 197-205.

Citing the ANNALS.	5 I. M. Smith, "On Regionalism and Common Sense," *Annals*, Association of American Geographers, Vol. 59 (1969), p. 4.

<table>
<tr><td>Repeat
citation.</td><td>6 <u>Ibid</u>. A frustrating entry which used the whole
line anyway. Make the citation: Smith, <u>op</u>. <u>cit</u>.,
footnote 5, p. 4. This is now standard <u>Annals</u>
style.</td></tr>
</table>

Handling several
authors and
varied topics
in a single footnote.
Do not give
separate numbers
to each author
when citing several
in one sentence,
but use one footnote
to cover them all.

7 Primary election problems were discussed by H. L. Long, "Problems of Primary Elections," <u>Political Scientist</u>, Vol. 9 (1965), pp. 1-13; the difficulties of getting vote totals during November ballotings were analyzed by J. B. Brown, "Counting the Vote," <u>Political Scientist</u>, Vol. 10 (1966), pp. 26-54; the stuffing of ballot boxes was reviewed by P. N. Slimm, "How Honest is Voting," <u>Political Scientist</u>, Vol. 8 (1964), pp. 98-107; and the issue of how many dead people vote was considered by J. J. White, "Some Dead Keep on Voting," <u>Political Scientist</u>, Vol. 11 (1967), pp. 238-66.

Citation forms for
nonrecurrent
reports and
irregular items.

8 R. U. Square, "The Analysis of Agricultural Regionalism," NRQ <u>Technical</u> <u>Report</u> <u>No. 1</u>, Project XZ-1111, 1964 (Washington, D. C.: Department of Agriculture, Bureau of Crops, 1965).

9 Government of India Planning Commission, <u>First Plan</u> (Dohra Dun, Uttar Pradesh, India: Government Printer, 1951), 721 pp.

10 U. C. Brown and I. C. Black, <u>Desert</u> <u>Water</u> <u>Supplies</u> <u>in</u> <u>the</u> <u>White</u> <u>Desert</u> <u>of</u> <u>Nevada</u> (Natick, Massachusetts: MQ Engineering and Research Office, U. S. Army, Technical Report EL-999, 1955), 201 pp.

Citation for
obscure items.

11 U. B. Bound, <u>Regionalism</u> <u>in</u> <u>Upper</u> <u>Pakistan</u> (published privately by the author through the Rawalpindi Bookmart, Rawalpindi, Pakistan, 1960), 99 pp.

12 C. U. Soon, <u>Regional</u> <u>Divisions</u> <u>of</u> <u>The Amazon</u> <u>Valley</u> (Aurora, Connecticut: Aurora Printers, for the River Explorers' Society, 1956), 101 pp.

Citing an article
published in an
edited book.

13 I. M. Brown, "When Models Become Problems," in B. A. Hoverstraw (Ed.), <u>The Making of Models</u> (New York: Mathematics Press, 1967), pp. 97-109.

Spacing of
initials.

14 Do not run initials together this way: H.C.G.St. Bernard. Space initials as though they were words, thus: H. C. G. St. Bernard; J. A. Jones, B. L. P. Smith, and W. H. Q. White.

FIGURE CAPTIONS FOR GEOGRAPHY AND GEOGRAPHERS

There must be an entry for each figure to be set by the printer, whether there is a caption or not. The period is placed after both the abbreviation and the number in all captions. Figure captions must be typed double spaced on a sheet, or sheets, separate from the text, footnotes, or the figures themselves.

Fig. 1.   The patterns of subregions within a major region.

Fig. 2.   (When an illustration, normally a self
            -titled map, is to be used and no expla-
            nation is needed, the figure number
            should be inserted and the space left
            blank; see illustration in Fig. 6, be-
            low.)

Fig. 3.   The details of regional subdivisions are formulated from specific data. Source: <u>United States Census</u> for 1930.

Fig. 4.   The basis for the regional boundary of the hill country can be seen clearly in the lower part of the photograph. (Photograph courtesy Greenland Travel Agency, New York, New York)

Fig. 5.   Diagram illustrating the overlapping of subregions.

Figure 6

Fig. 7.   The spatial pattern of the referendum. Source: Yarrow, <u>op. cit.</u>, footnote 16, map on p. 43.

APPENDIX III

Dissertations and Theses

Whittlesey, Derwent, "Dissertations in Geography Accepted by Universities in the United States for the Degree of Ph.D. as of May, 1935," *Annals*, Association of American Geographers, Vol. 25, No. 4 (December, 1935), pp. 211–37.

Hewes, Leslie, "Dissertations in Geography Accepted by Universities in the United States for the Degree of Ph.D., 1935–June, 1946," *Annals*, Association of American Geographers, Vol. 36, No. 4 (December, 1946), 215–47.

Hewes, Leslie, "Recent Geography Dissertations and Theses Completed and in Preparation," *The Professional Geographer* (title varies).

Vol. II
  (January, 1950), 8–18
  (March, 1950), 11–20
  (June, 1950), 14–20
  (October, 1950), 11–14
Vol. III
  (February, 1951), 10–17
  (July, 1951), 12–15
  (November, 1951), 34–39
Vol. IV
  (July, 1952), 36–39
  (November, 1952), 39–45

Vol. V
  (May, 1953), 34–45
  (November, 1953), 50–55
Vol. VI
  (March, 1954), 55–57
  (November, 1954), 21–29
Vol. VII
  (March, 1955), 26–27
  (November, 1955), 22–28
Vol. VIII
  (March, 1956), 42–45
  (November, 1956), 26–32

Vol. IX
   (March, 1957), 51–54
   (November, 1957), 29–33
Vol. X
   (March, 1958), 29–34
   (November, 1958), 33–41
Vol. XI
   (January, 1959,
      Part II), 134–45
Vol. XII
   (March, 1960), 24–26
   (November, 1960), 23–30
Vol. XIII
   (November, 1961), 52–58
Vol. XIV
   (November, 1962), 34–42
Vol. XV
   (November, 1963), 35–47

Vol. XVI
   (November, 1964), 31–43
Vol. XVII
   (November, 1965), 50–62
Vol. XVIII
   (November, 1966), 387–
      401
Vol. XIX
   (November, 1967), 359–
      73
Vol. XX
   (November, 1968), 423–
      44
Vol. XXI
   (November, 1969), 419–
      44

PART II

AIDS TO GEOGRAPHICAL RESEARCH

CHAPTER 5

General Guides, Bibliographies, and Sources of Information

A. Guides to the Use of Reference Materials

1. ALEXANDER, CARTER, and ARRID J. BURKE, *How to Locate Educational Information and Data*, New York: Bureau of Publications, Columbia University, 1958. 419 pp.

2. BARTON, MARY H. (comp.), *Reference Books, A Brief Guide for Students and Other Users of the Library*, 4th ed., Baltimore: Enoch Pratt Free Library, 1959. 117 pp.

3. CHANDLER, GEORGE, *How to Find Out; A Guide to Sources of Information for All*, 2nd ed., Oxford: Pergamon Press, 1966. 198 pp. (A source guide for material classified under the Dewey decimal system.)

4. DOWNS, ROBERT B., *How to Do Library Research*, Urbana: University of Illinois Press, 1966. 179 pp.

5. HIRSHBERG, HERBERT S., and CARL MELNAT, *Subject Guide to Reference Books*, Chicago: American Library Association, 1942. 228 pp.

6. LAZARSFELD, PAUL F., *The Language of Social Research*, Glencoe, Ill.: Free Press, 1955. 590 pp.

7. LINDEN, RONALD O., *Books and Libraries: A Guide for Students*, London: Cassell, 1965. 308 pp.

8. MALCLES, LOUISE-HOELLE, *Les Sources du Travail Bibliographique*, Paris: 1950–1958. 3 vols. in 4. (Lists the bibliographies and reference works for each subject as well as the standard histories and introductory materials; reprinted in 1965.)

9. MURPHEY, ROBERT, *How and Where to Look It Up: A Guide to Standard Sources of Information*, New York: McGraw-Hill, 1958. Section 15, "How to Find Out About Places," pp. 259–310; "Geography," p. 436.

10. PARADIS, ADRIAN, *The Research Handbook; A Guide to Reference Sources*, New York: Funk & Wagnalls, 1966. 217 pp.

11. SABOR, JOSEFA E., *Manual de fuentes de información; obras de referencia: enciclopedias, diccionarios, bibliografías, biografías, etc.*

12. SHORES, LOUIS, *Basic Reference Sources: Introduction to Materials and Methods*, Chicago: American Library Association, 1954.

13. WALFORD, A. J., with the assistance of L. M. PAYNE (eds.), *Guide to Reference Material*, London Library Association, 1959. Supplement, "Geography, Exploration, Travel," 1963, pp. 392–408. (2nd edition, 1966.)

14. WHITE, CARL M., and others, *Sources of Information in the Social Sciences*, Totowa, N.J.: Bedminster Press, 1964. 498 pp.

15. WILSON, E. BRIGHT, *An Introduction to Scientific Research*, New York: McGraw-Hill, 1952. 375 pp.

16. WINCHELL, CONSTANCE M., *Guide to Reference Books*, 8th ed., Chicago: American Library Association, 1967. 741 pp. and Supplements, 1950 to date. Section U–Geography, pp. 459–475; Supplement 1950–52, Geography, pp. 79–81; Second Supplement 1953–55, Geography, pp. 90–93; Third Supplement 1956–58, pp. 92–95.

17. WRIGHT, JOHN K., and ELIZABETH T. PLATT, *Aids to Geographical Research*, 2nd ed., New York: American Geographical Society, 1947. 331 pp.

B. Bibliographies and Card Catalogs

18. AMERICAN GEOGRAPHICAL SOCIETY OF NEW YORK, *Research Catalogue*, New York: 1962. 15 vols. and map supplement.

19. BESTERMAN, THEODORE, *A World Bibliography of Bibliographical Catalogues, Calendars, Abstracts, Digests, Indexes and the Like*, 3rd ed., Geneva: Societas Bibliographica, 1955–58. 4 vols. 4th ed., 1965. 6 vols. (Geography—Vol. II, pp. 2436–2457).

20. *Bibliographic Index: A Cumulative Bibliography of Bibliographies*, New York: H. W. Wilson.

21. BISHOP, BERNICE P., Museum, Honolulu, Hawaii, *Dictionary Catalog of the Library*, Boston: G. K. Hall, 1964. 9 vols.

22. *Book Review Digest*, New York: H. W. Wilson.

23. *British Books in Print*, New York: R. R. Bowker. Quadrennially.

24. *The British National Bibliography Cumulated Index 1955–1959*, London: The Council of the British National Bibliography, Ltd., 1961. 3 vols. 1950–1954 also available. Cumulation published every 3, 6, 9, 12 months on books published in England.

25. CALIFORNIA, UNIVERSITY OF, BERKELEY LIBRARY, *Author–Title Catalog*, 1963–64. 115 vols.

26. CALIFORNIA, UNIVERSITY OF, LOS ANGELES LIBRARY, *Dictionary Catalog of the University Library 1919–1962*, 1963. 129 vols.

27. *Catalog of the United States Geological Survey Library*, Boston: G. K. Hall, 1966. 25 vols.

28. DARTMOUTH COLLEGE LIBRARY, *Dictionary Catalog of the Stefansson Collection on the Polar Regions*, Boston: G. K. Hall. 8 vols.

29. DOWNS, ROBERT B., and FRANCES B. JENKINS (eds.), "Bibliography: Current State and Future Trends," *Library Trends*, 15 (March and April, 1967), 337–908.

30. HARVARD UNIVERSITY, PEABODY MUSEUM OF ARCHAEOLOGY AND ETHNOLOGY, *Author and Subject Catalogues*, Boston: G. K. Hall, 1963. 53 vols.

31. HARVEY, ANTHONY P., *Directory of Scientific Directories*, Hodgson, Guernsey, Channel Islands, Great Britain: 1969. 272 pp.

32. *An Index to Book Reviews in the Humanities*, Detroit: P. Thomson, 1960. Quarterly, with annual cumulation.

33. KELLY, JAMES, *The American Catalogue of Books Published in the United States from January, 1861, to January, 1871*, New York: Peter Smith, 1938. 2 vols.

34. NEW YORK PUBLIC LIBRARY, REFERENCE DEPARTMENT, *Dictionary Catalog of the History of the Americas*, New York: 1961. 28 vols.

35. PEDDIE, R. A., *Subject Index of Books Published up to and Including 1880*, London: H. Pordes, 1962. 745 pp.

36. *Subject Guide to Books in Print: An Index to the Publisher's Trade List Annual*, New York: R. R. Bowker, 1956 to date. Annual.

37. *Technical Book Review Index*, New York: Special Libraries Association, 1917 to date.

38. *United States Catalog: Books in Print, January 1, 1928*, New York: H. W. Wilson, 1912–1928. *Cumulative Book Index: A World List of Books in the English Language*, New York: H. W. Wilson, 1928 to date.

39. U.S. GEOLOGICAL SURVEY LIBRARY, *Catalog*, Boston: G. K. Hall, 1964. 25 vols.

40. U.S. LIBRARY OF CONGRESS, *A Catalog of Books Represented by the Library of Congress Printed Cards*, Ann Arbor, Mich.: Edwards Brothers, 1942–46. 167 vols.

C. Government Documents and Their Use

41. ANDROIT, JOHN L., *Guide to U.S. Government Serials and Periodicals*, 1964 ed., McLean, Va.: Documents Index 1964. 3 vols.

42. BODY, ALEXANDER, *Annotated Bibliography of Bibliographies on Selected Government Publications and Supplementary Guides to the Superintendent of Documents Classification System*, Kalamazoo: Western Michigan University, 1967. 181 pp.

43. BOYD, ANNE M., *United States Government Publications*, 3rd ed., New York: H. W. Wilson, 1949.

44. BRIMMER, BRENDA, et al, *A Guide to the Use of United Nations*

Documents, Dobbs Ferry, N.Y.: Occana Publications, 1962. 272 pp.

45. BROWN, EVERETT S., *Manual of Government Publications United States and Foreign,* New York: Appleton–Century–Crofts, 1950.

46. *California State Publications,* Sacramento: Printing Division, Documents Section, 1947.

47. CHILDS, J. B., *Government Documents Bibliography in the United States and Elsewhere,* 3rd ed., Washington: 1942.

48. GREGORY, WINIFRED, *List of Serial Publications of Foreign Governments, 1815–1931,* New York: H. W. Wilson, 1932.

49. HIRSHBERG, HERBERT S., and CARL MELINOT, *Subject Guide to United States Government Publications,* Chicago: American Library Association, 1947. 228 pp.

50. JACKSON, ELLEN, *Subject Guide to Major U.S. Government Publications,* Chicago: American Library Association, 1968.

57. POORE, BEN P., *A Descriptive Catalog of the Government Publications of the United States, Sept. 5, 1774–March 4, 1881,* Washington: 1885. 1,392 pp.

52. RICKLES, ROBERT, *Marketing Guide to U.S. Government Research and Development,* Park Ridge, N.J.: Hoyes Development Corp., 1966. 229 pp.

53. SCHMECKEBIER, LAWRENCE F., *Government Publications and Their Use,* Washington: The Brookings Institution, 2nd rev. ed., 1969.

54. *United Nations Documents Index,* New York. Monthly since January, 1950.

55. *U.S. Congressional Directory,* Washington. Annual. (Lists governmental officials; gives addresses; includes maps of congressional districts.)

56. U.S. DEPARTMENT OF COMMERCE, BUREAU OF THE CENSUS, *Bureau of the Census—Fact Finder for the Nation,* Washington: 1965. 59 pp.

57. ———, *Guide to Census Bureau Statistics: Subjects and Areas,* Washington: 1963. 47 pp.

58. ———, *Publications of Foreign Countries, An Annotated Accessions List Prepared for Official Use,* Washington. Quarterly since 1947.

59. U.S. DEPARTMENT OF COMMERCE, NATIONAL BUREAU OF STANDARDS, INSTITUTE FOR APPLIED TECHNOLOGY, *Government-Wide Index to Federal Research and Development Reports*, Washington. Irregularly.

60. ———, *U.S. Government Research and Development Reports*, Washington. Irregularly.

61. U.S. DEPARTMENT OF STATE, OFFICE OF EXTERNAL RESEARCH, *Governmental Resources Available for Foreign Affairs Research*, Linda Lowenstein, Washington: 1967. 83 pp.

62. *United States Government Publications: Monthly Catalog*, Washington: Government Printing Office, 1895 to date.

63. U.S. LIBRARY OF CONGRESS, *A Directory of Information Resources in the United States*, Federal Government, Washington: 1967. 419 pp.

64. U.S. LIBRARY OF CONGRESS, CENSUS LIBRARY PROJECT, *Catalog of United States Census Publications 1790–1945*, prepared by Henry J. Dubester, Washington: 1950.

65. ———, *State Censuses: An Annotated Bibliography of Censuses of Population Taken After the Year 1790 by States and Territories of the United States*, prepared by Henry J. Dubester, Washington: Government Printing Office, 1948. 73 pp.

66. ———, *Statistical Bulletins, An Annotated Bibliography of the General Statistical Yearbooks of the Major Political Subdivisions of the World*, prepared by Phyllis G. Carter, Washington: 1954.

67. U.S. LIBRARY OF CONGRESS, EXCHANGES AND GIFT DIVISION, *Monthly Checklist of State Publications*, Washington: 1910 to date.

68. U.S. NATIONAL ARCHIVES, DIVISION OF THE FEDERAL REGISTER, *United States Government Organization Manual*, Washington. Annually.

69. U.S. SUPERINTENDENT OF DOCUMENTS, *Catalogue of the Public Documents of Congress and of All Departments of the Government of the United States for the Period March 4, 1893–December 31, 1940*, Washington: 1896–1945. 25 vols.

70. VINGE, CLARENCE L. and GRACE A., *U.S. Government Publications for Teaching and Research in Geography*, Norman, Okla.: National Council for Geographic Education, 1962. 105 pp.

71. ———, *U.S. Government Publications for Teaching and Research in Geography and Related Disciplines*, Totowa, N.J.: Littlefield, Adams and Co., 1966.

72. WILCOX, JEROME K., *Bibliography of New Guides and Aids to Public Documents Use, 1953–1956*, New York: Special Libraries Association, 1957. 16 pp.

73. ———, *Manual on the Use of State Publications*, Chicago: American Library Association, 1940.

D. Serial Listings and Guides to Periodical Literature

74. Ayer, N. W., & Sons Directory, *Newspapers and Periodicals*, Philadelphia. Annual.

75. BOALCH, DONALD H., *Current Agricultural Serials: A World List of Serials in Agriculture and Related Subjects*, Oxford: International Association of Agricultural Librarianists and Documentalists, 1965. 351 pp.

76. BOEHM, ERIC, and LALIT ADOLPHUR, *Historical Periodicals: An Annotated World List of Historical and Related Serial Publications*, Santa Barbara: Clio Press, 1961. 618 pp.

77. CANADIAN LIBRARY ASSOCIATION, *Canadian Index, A Guide to Canadian Periodicals and Documentary Films*, Ottawa: 1947 to date. Monthly guide to over 60 Canadian periodicals.

78. CUSHING, HELEN G., and ADAH V. MORRIS, *Nineteenth Century Readers' Guide to Periodical Literature, 1890–1899, with Supplementary Indexing, 1900–1922*, New York: H. W. Wilson, 1944. 2 vols.

79. *Directory of Canadian Scientific and Technical Periodicals*, Ottawa: National Research Council, 1961.

80. HARRIS, CHAUNCY D., *Annotated World List of Selected Current Geographical Serials in English*, 2nd ed., rev., University of Chicago, Department of Geography, Research Paper No. 96, Chicago: 1964. 32 pp.

81. ———, "Geographical Serials of Latin America," *Revista Geografica* (June, 1966), 148–167.

82. ———, and JEROME D. FELLMAN, *International List of Geographical Serials*, University of Chicago, Department of Geography, Research Paper No. 63, Chicago: 1960. 194 pp.

83. *Irregular Serials and Annuals; An International Directory*, New York: R. R. Bowker, 1967. 668 pp.

84. *The New York Times Index*, New York: The New York Times, 1913 to date.

85. PAN AMERICAN UNION, *Directory of Current Latin American Periodicals*, UNESCO bibliographical handbook No. 8, Paris: UNESCO, 1958. 266 pp.

86. POOLE, WILLIAM F., WILLIAM I. FLETCHER, and MARY POOLE (eds.), *Poole's Index to Periodical Literature 1802–81*, and supplementary vols. to 1906, Gloucester, Mass.: Peter Smith. Various dates.

87. *Readers' Guide to Periodical Literature, Nineteenth Century Readers' Guide to Periodical Literature, 1890–1899, with Supplementary Indexing, 1900–1922*, 2 vols.

88. *Readers' Guide to Periodical Literature*, 1901 to date.

89. *Social Science and Humanities Index* (formerly International Index), 1908 to date.

90. *Standard Periodical Directory*, 1967 ed., New York: Oxbridge Publishing Co., 1966. 1019 pp.

91. The *Times*, London, *Palmer's Index to the Times Newspaper, 1790–1905.*

92. The *Times*, London, *Index to the Times*, 1914 to date.

93. *Ulrich's Periodicals Directory. A Classified Guide to a Selected List of Current Periodicals, Foreign and Domestic*, 13th ed., 1969.

94. *Union List of Serials in Libraries of the United States*, 3rd ed., New York: H. W. Wilson, 1965. Kept up to date by *New Serial Titles; A Union List of Serials*.

95. U.S. LIBRARY OF CONGRESS, REFERENCE DEPARTMENT, *Serials for African Studies*, Helen F. Conover (comp.), Washington: 1961. 163 pp.

96. *Wall Street Journal, Index*, New York: 1958 to date. Monthly, with annual cumulations.

E. Miscellaneous Materials

97. ALEXANDER, RAPHAEL (ed.), *Sources of Information and Unusual Services*, 7th ed., New York: International Directory Co., 1964. 100 pp.

98. BLACK, DOROTHY M., *Guide to Lists of Master's Theses*, Chicago: American Library Association, 1965. 144 pp.

99. CHIDDELL, P. R., "Bibliography of Bibliographies of Free and Inexpensive Learning Aids," *Curriculum Bulletin*, 17 (October, 1961), 216.

100. DENVER, ESTHER, *Sources of Free and Inexpensive Educational Materials*, 2nd ed., rev., Grafton, W. Va.: The Author, 1963. 400 pp.

101. *The Directory of the Forest Products Industry*, Portland, Ore.: Miller Freeman Publications, 1962 to date. Annual.

102. *Directory of National Trade and Professional Associations of the United States*, Washington: Potomac Books, 1966.

103. *Dissertation Abstracts: Abstracts of Dissertations and Monographs in Microform*, Ann Arbor, Mich.: University Microfilms, 1961 to date. Similar abstracts for earlier years published as: *Dissertation Abstracts: A Guide to Dissertations and Monographs Available in Microform* (1951–62); *Microfilm Abstracts: A Collection of Doctoral Dissertations and Monographs Available in Complete Form on Microfilm* (1938–51); *Doctoral Dissertations Accepted by American Universities* (1933–55).

104. DRAKE, MILTON, *Almanacs of the United States*, New York: Scarecrow Press, 1962. 2 vols.

105. *The Foundation Directory*, 3rd ed., New York: Russell Sage, 1967. 1198 pp.

106. Fox, WILLIAM L., *List of Doctoral Dissertations in History in Progress or Completed at Colleges and Universities in the United States since 1958*, Washington: The American Historical Association, 1961. 61 pp.

107. GABLER, ROBERT E. (ed.), *A Handbook for Geography Teachers,* Normal, Ill.: National Council for Geographic Education, 1966. 273 pp.

108. GEORGE PEABODY COLLEGE FOR TEACHERS, DIVISION OF SURVEYS AND FIELD SERVICES, *Free and Inexpensive Learning Materials,* Nashville, Tenn.: 1941 to date. A biennial listing.

109. GOGGIN, DANIEL T. (comp.), *Preliminary Inventory of the Records Relating to International Boundaries* (Record Group 76), Washington: National Archives and Records Service, 1968. 98 pp.

110. *Guide to American Directories,* Englewood Cliffs, N.J.: Prentice-Hall, 1965.

111. HODGSON, JAMES GOODWIN, *The Official Publications of American Counties, A Union List,* Fort Collins, Colo.: 1937. (University Microfilms, 1966 reprint.)

112. KRUZAS, ANTHONY T., *Directory of Special Libraries and Information Services,* Detroit: Gale Research, 1963. 767 pp.

113. *Museums Directory of the United States and Canada,* 2nd ed., Washington: American Association of Museums, 1965. 1039 pp.

114. NATIONAL RESEARCH COUNCIL, *Scientific and Technical Societies of the United States and Canada,* 7th ed., Washington: 1961. Parts 1, 2.

115. PEPE, THOMAS J., *Free and Inexpensive Educational Aids,* New York: Dover Publications, 1960. 289 pp.

116. *Research Centers Directory,* Detroit: Gale Research, 1960. (Kept up-to-date by the *Quarterly,* New Research Centers.)

117. RUFFNER, FREDERICK G., JR. (ed.), and others, *Encyclopedia of Associations,* 4th ed., Detroit: Gale Research, 1964. 3 vols. (Kept up-to-date by periodic supplements.)

118. SCHAIN, ROBERT L., and MURRAY POLNER, *Where to Get and How to Use Free and Inexpensive Teaching Aids,* New York: Teachers Practical Press, 1963. 63 pp.

119. *Sources of Free and Inexpensive Educational Materials,* Chicago: Field Enterprises, 1958.

120. U.S. DEPARTMENT OF COMMERCE, OFFICE OF TECHNICAL SERVICES, *Directory of National Associations of Businessmen,* Jay Judkins, Washington: 1961. 81 pp.

121. WAGNER, GUY, AND EDNA CHRISTOPHEL, *Free Learning Materials for Classroom Use*, Cedar Falls, Iowa: State College of Iowa, 1963. 62 pp.

122. WASSERMAN, PAUL, *Information for Administrators, A Guide to Publications and Services in Business and Government*, Ithaca: Cornell University Press, 1956. 375 pp.

123. *The World of Learning*, London: Europa Publications, 1947 to date. Annual guide; gives names and addresses of world organizations.

CHAPTER 6

Special Indexes, Abstracts, and Bibliographies

A. Geographical

124. AKADEMIIA NAUK SSSR: INSTITUTE NAUCHNOG INFORMATSII, *Referatiunyy Zhurnal Geografiia*, Moscow: 1956. (Monthly, annotated bibliography with titles in the language in which the article appeared originally.)

125. AMERICAN GEOGRAPHICAL SOCIETY, *Current Geographical Publications: Additions to the Research Catalogue of the American Geographical Society*, Vol. I, New York: 1938 to date. Monthly except July and August.

126. ANDERSON, MARC, *A Working Bibliography of Mathematical Geography*, Michigan Inter-University Community of Mathematical Geographers, Discussion Paper No. 2, Ann Arbor: University of Michigan, Department of Geography, 1963. 52 pp.

127. ASSOCIATION DE GÉOGRAPHES FRANÇAIS, *Bibliographie Géographique Internationale*, Paris: Armand Colin, 1891–1953; Centre National de la Recherche Scientifique, 1954 to date. Annually since 1922.

128. ASSOCIATION OF AMERICAN GEOGRAPHERS, COMMISSION ON COLLEGE GEOGRAPHY, *A Geographical Bibliography for American Col-*

lege Libraries, Pub. No. 9, Gordon R. Lewthwaite, Edward T. Price, Jr., and Harold A. Winters (comps. and eds.), Washington: 1970.

129. BERGEN, J. V., *Introductory College Geography Textbooks, 1940–1965; A Bibliographical Summary,* Normal, Ill.: National Council for Geographic Education, 1966.

130. BERRY, BRIAN, and ALLEN PRED, *Central Place Studies: A Bibliography of Theory and Applications,* Regional Science Research Institute, Bibliography Series, No. 1, Philadelphia: Regional Science Research Institute, 1961. 153 pp. Supplement through 1964, 1965. 50 pp.

131. *Bibliographie Cartographique Internationale,* Paris: Armand Colin. Annual since 1948.

132. BURKETT, JACK (ed.), *Concise Guide to the Literature of Geography,* Ealing, England: Ealing Technical College, 1967. 47 pp.

133. CONS, G. J., and R. C. HONEYBONE (eds.), *Handbook for Geography Teachers,* London: Methuen, 1960. 525 pp.

134. "Cumulative Index 1946–1966," *The Professional Geographer,* 19 (March, 1967), 65–116.

135. DURRENBERGER, ROBERT W., *Environment and Man: A Bibliography,* Palo Alto: National Press Books, 1970.

136. FELLAND, NORDIS A., "Geography," *Library Trends,* 15 (April, 1967), 704–709.

137. *Geographisches Jahrbuch,* Gotha: VEB Herman Haack Geographisch-Kartographische Anstalt, 1866–1956. 61 vols.; none since 1956.

138. *Geomorphological Abstracts,* London: Department of Geography, London School of Economics, 1960 to date.

139. GROTEWOLD, ANDREAS, *A Selective Annotated Bibliography of Publications Relevant to the Geographical Study of International Trade,* Columbia: University of Missouri, Department of Geography, 1960.

140. HARRIS, CHAUNCY D., *A Bibliography of the Geography of Manufacturing,* Chicago: University of Chicago, Department of Geography, 1952. 26 pp.

141. HORNSTEIN, H. A., *A Bibliography of Paperback Books Relating*

to Geography, Chicago: National Council for Geographic Education, 1970.

142. Lock, C. B. Muriel, *Geography: A Reference Handbook*, London: Clive Bingley, 1968. 179 pp.

143. London, University of, Institute of Education, *Handbook for Geography Teachers*, 5th ed., M. Long (ed.), London: Methuen, 1964. 534 pp.

144. Minto, C. S., *How to Find Out in Geography*, Oxford: Pergammon Press, 1966. 99 pp. (A source guide for material classified under the Dewey decimal system.)

145. Norell, Irene P., *Geographical Literature: A Brief Annotated Guide*, San Jose, Calif.: The Author, 1969.

146. Peltier, Louis C., *Bibliography of Military Geography*, Washington: Association of American Geographers, 1962. 76 pp.

147. Polska Akademia Nauk, *Polska Bibliografia Analityczna*, Warsaw: 1956 to date.

148. Royal Geographical Society, *New Geographical Literature and Maps*, London: 1951; June, 1966.

149. Siddall, William R., *Transportation Geography: A Bibliography*, Bibliography Series No. 1, Manhattan, Kans.: Kansas State University Library, 1964. 46 pp.

150. Sommer, John W., *Bibliography of Urban Geography, 1940–1964*, Geographical Publications at Dartmouth No. 5, Hanover, N.H.: 1966. 92 pp.

151. United Nations Educational, Scientific, and Cultural Organization, *Source Book for Geography Teaching*, London: Longmans, 1965. 255 pp. (Sources—pp. 201–255.)

152. U.S. Library of Congress, Map Division, *A List of Geographical Atlases in the Library of Congress*, Clara E. Le Gear (comp.), Washington: 1958 to date.

153. *Westermanns Geographische Bibliographie*, Braunschweig: Georg Westermann Verlag, 1955 to date. (10 issues per year.)

154. Wolfe, Roy I., *An Annotated Bibliography of the Geography of Transportation*, Institute of Transportation and Traffic Engineering, Information Circular No. 29, Berkeley: University of California Press, 1961. 61 pp.

B. Physical, Biological, and Earth Sciences

155. ABELL, L. F., and W. J. GELDERMAN (comps.), *Annotated Bibliography on Reclamation and Improvement of Saline and Alkali Soils, 1957–1964,* Wageningen, Netherlands: International Institute for Land Reclamation and Improvement, 1964. 59 pp.

156. AGERTER, SHARLENE, and WALDO S. GLOCK, *Annotated Bibliography of Tree Growth and Growth Rings, 1950–1962,* Tucson: University of Arizona, 1964. 188 pp.

157. AMERICAN CHEMICAL SOCIETY, *Chemical Abstracts,* Easton, Pa.: 1907 to date.

158. AMERICAN INSTITUTE OF CHEMICAL ENGINEERS, WATER COMMITTEE, *A Bibliography of Books on the Environment—Air, Water and Solid Wastes,* Toledo, Ohio: University of Toledo, 1964. 54 pp.

159. AMERICAN METEOROLOGICAL SOCIETY, *Bibliography on Physical Oceanography,* Boston: 1970.

160. ———, *Bibliography on Physical Oceanography of the Indian Ocean,* Boston: 1966.

161. ———, *Collected Bibliographies on Physical Oceanography* (1953–1964), Boston: 1966.

162. ———, *Cumulative List—Translations of Foreign Literature Relating to the Atmospheric Sciences, 1952–1964,* Boston: 1966.

163. ANDERSONY, FERENC (comp.), "Smog and Los Angeles," *Bulletin,* Special Libraries Association, Geography and Map Division, 70 (December, 1967), 11–15.

164. *Annotated Bibliography of Economic Geology,* Urbana, Ill.: Economic Geology Publishing Co., 1928 to date. Semiannual.

165. *Annotated Bibliography on Hydrology,* Washington: 1935 to date. (Published in various years by American Geophysical Union, U.S. Geological Survey, U.S. Inter-Agency Committee on Water Resources.)

166. *Applied Science & Technology Index*, New York: H. W. Wilson. Annually since 1958. Formerly, *The Industrial Arts Index* (1913–1957).

167. THE BIBLIOGRAPHICAL CENTER FOR RESEARCH ROCKY MOUNTAIN REGION, *Reclamation 1902–1938: A Supplemental Bibliography*, John G. Gaul (comp.), Denver: 1939. 98 pp.

168. *Bibliography and Index of Geology Exclusive of North America*, New York: Geological Society of America, 1930 to date. Annual 1933–1965; monthly abstracts, annual index since 1965.

169. *Biological Abstracts*, Philadelphia: BioSciences Information Service. Bimonthly with annual cumulation since 1927.

170. *Biological and Agricultural Index*, New York: H. W. Wilson, 1919 to date. Formerly *Agricultural Index*.

171. BLANCK, FRED C., *Handbook of Food and Agriculture*, New York: Reinhold, 1955. (Lists of research groups, scientific societies, trade associations, and agricultural experiment stations.)

172. BRITISH SOCIETY FOR RESEARCH IN AGRICULTURAL ENGINEERING, *Agricultural and Horticultural Engineering Abstracts*, Wrest Park, Silsoe, England: 1950. Quarterly.

173. BUREAU DE RECHERCHES GÉOLOGIQUES ET MINIÈRES, *Guide Bibliographique d'hydrogéologie*, Paris: 1964. 113 pp.

174. CALIFORNIA STATE DEPARTMENT OF WATER RESOURCES, *Index to Publications, State Department of Water Resources*, Sacramento: 1965. 40 pp.

175. ———, *A Survey of Water Resources Information and Information Repositories in California*, Sacramento: 1964. 94 pp.

176. CALIFORNIA, UNIVERSITY OF, BUREAU OF PUBLIC ADMINISTRATION, *Water Plans for California: A Bibliography*, Dorothy C. Tompkins (comp.), Berkeley: 1961. 180 pp.

177. CALIFORNIA, UNIVERSITY OF, ENGINEERING RESEARCH, *Saline Water Demineralization—A Review and Bibliography*, J. W. McCutchan, Los Angeles: 1961. 94 pp.

178. CALIFORNIA, UNIVERSITY OF, WATER RESOURCES CENTER ARCHIVES, *Bachelor of Science Theses on Water Resources Engineering*, Gerald J. Giefer and Cynthia Barnes (comps.), Report No. 6, Berkeley: 1959. 93 pp.

179. ———, *A Bibliography of the Reports and Publications of James Dix Schuyler*, Lois Judd (comp.), Report No. 11, Berkeley: 1960, 60 pp.

180. ———, *Bulletins and Reports of California State Water Agencies*, Report No. 15, Berkeley: 1963. 157 pp.

181. ———, *Index to Periodical Literature on Aspects of Water in California*, Cynthia Barnes and Gerald J. Giefer (comps.), Report No. 14, Berkeley: 1963. 337 pp.

182. ———, *Publications and Reports of Charles Gilman Hyde*, Gerald J. Giefer, Cynthia Barnes, Morrill G. Folsom (comps.), Report No. 4, Berkeley: 1959. 38 pp.

183. ———, *Reports and Data in the Water Resources Collection, Honnold Library, Claremont Colleges, Claremont, California*, Gerald J. Giefer and Cynthia Barnes (comps.), Report No. 10, Berkeley: 1960. 66 pp.

184. ———, *Theses on Engineering, Economic, Social and Legal Aspects of Water*, Emily C. Lumbard (comp.), Report No. 2, Berkeley: 1958. 73 pp.

185. ———, *Theses on Water Resources, Stanford University, California Institute of Technology and University of Southern California*, Gerald J. Giefer, Lois Judd, and Cynthia Barnes (comps.), Report No. 7, Berkeley: 1959. 81 pp.

186. ———, *Water Pollution Data, Regional Water Pollution Control Boards, State of California*, Gerald J. Giefer and Cynthia Barnes (comps.), Report No. 5, Berkeley: 1959. 79 pp.

187. ———, *Water Resources Reports and Data in the Bernard A. Etcheverry Collection*, Morrill G. Folsom and Wilma J. Woodward (comps.), Report No. 1, Berkeley: 1958. 170 pp.

188. ———, *Water Resources Reports and Data, Los Angeles County Flood Control District*, Cynthia Barnes (comp.), Report No. 9, Berkeley: 1960. 77 pp.

189. ———, *Water Resources Reports by Walter Leroy Huber*, Lois Judd and Clare Bullitt (comps.), Report No. 12, Berkeley: 1962. 103 pp.

190. ———, *Watershed Management Research Data, U.S. Department of Agriculture, Forest Service, California Forest and Range*

137

Experiment Station, Berkeley and Glendora, California, Cynthia Barnes and Morrill G. Folsom (comps.), Report No. 3, Berkeley: 1959. 127 pp.

191. ———, *Water Wells, An Annotated Bibliography,* Report No. 3, Berkeley: 1963. 141 pp.

192. CANADA, GEOLOGICAL SURVEY OF, *Index of Publications of Geological Survey of Canada,* Ottawa. Annual.

193. CHRONIC, JOHN and HALKA, *Bibliography of Theses in Geology, 1958–1963,* Washington: American Geological Institute, 1965. 268 pp.

194. CLARK, ALAN W., *Dams, A Bibliography,* Fort Belvoir, Va.: The Engineer School, 1936. 256 pp.

195. COMMONWEALTH BUREAU OF SOIL SCIENCE, *Bibliography of Soil Science, Fertilizers and General Agronomy,* Farnham Royal Bucks, Great Britain: 1931–1962. 8 vols. to date.

196. CULVER, WANE E., *Effects of Cold on Man; An Annotated Bibliography,* Washington: American Physiological Society, 1959.

197. EARTH SCIENCE CURRICULUM PROJECT, *Free Material for Earth Science Teachers,* Englewood Cliffs, N.J.: Prentice-Hall, 1965. 24 pp.

198. ———, *Planetariums, Observatories, and Earth Science Exhibits,* Englewood Cliffs, N.J.: Prentice-Hall, 1965. 32 pp.

199. ———, *Sources of Earth Science Information,* Englewood Cliffs, N.J.: Prentice-Hall, 1964. 44 pp.

200. *Engineering Index,* New York: American Society of Mechanical Engineers. Published since 1892.

201. FRITZ, EMANUEL, *California Coast Redwood [Sequoia sempervirens (D. Don) Endl.], An Annotated Bibliography to and including 1955,* San Francisco: Foundation for American Resource Management, 1957. 267 pp.

202. FRY, BERNARD M., and FOSTER E. MOHRHARDT (eds.), *A Guide to Information Sources in Space Science and Technology,* New York: Interscience Publishers, 1963. 579 pp.

203. *Geoscience Abstracts,* Washington: American Geological Institute. Monthly. 1959–1966 inclusive material to be included in *Abstracts of North American Geology.*

204. *Geoscience Documentation*, London: Lea Associates, 1969 to date. Monthly.

205. HOWELL, J. V., and A. I. LEVORSEN, *Directory of Geological Material in North America*, Washington: American Geological Institute, 1957. 208 pp.

206. INTERAGENCY COMMITTEE ON OCEANOGRAPHY, *Bibliography of Oceanographic Publications*, Pamphlet No. 9, Washington: 1963. 23 pp.

207. INTERNATIONAL COMMISSION ON IRRIGATION AND DRAINAGE, *Bibliography on Irrigation, Drainage, River Training and Flood Control*, New Delhi: 1960. 60 pp.

208. JACOBSTEIN, J. MYRON, and ROY M. MERSKY, *Water Law Bibliography, 1847–1965*, Silver Spring, Md.: Jefferson Law Book Co., 1966. 250 pp.

209. KAPLAN, STUART R., *Guide to Information Sources in Mining, Minerals, and Geosciences*, Vol. II of *Guides to Information Sources in Science and Technology*, New York: Interscience Publishers, 1955. 599 pp.

210. KNOBBE, MARY L., *Air Pollution: A Non-Technical Bibliography (Annotated)*, Monticello, Ill.: Council of Planning Librarians, 1969. 9 pp.

211. KRAMER, H. P., "Selective Annotated Bibliography on Tornadoes," *Meterological Abstracts*, 1 (1950), 307–332.

212. KRAMER, MOLLIE P., "Annotated Bibliography on Storm Surges," *Meteorological Abstracts*, 6 (1955), 370–392.

213. MASON, BRIAN, *The Literature of Geology*, New York: 1953. 155 pp.

214. *McGraw-Hill Basic Bibliography of Science and Technology*, New York: McGraw-Hill, 1966. 738 pp.

215. *Meteorological and Geoastrophysical Abstracts*, formerly *Meteorological Abstracts* (1950–59), Boston: American Meteorological Society, 1950 to date.

216. MICHIGAN UNIVERSITY INSTITUTE FOR COMMUNITY DEVELOPMENT, *An Annotated Bibliography on Water Problems*, Technical Bulletin B-29, East Lansing: 1962. 23 pp.

217. MINERALOGICAL SOCIETY OF GREAT BRITAIN, *Mineralogical Abstracts*, London: 1959 to date. Quarterly.

218. MITCHELL, JAMES K., *A Selected Bibliography of Coastal Erosion, Protection and Related Human Activity in North America and the British Isles*, Chicago: Department of Geography, University of Chicago, 1968. 66 pp.

219. NATIONAL ACADEMY OF SCIENCES—NATIONAL RESEARCH COUNCIL, *Oceanography Information Sources*, Publication 1417, Washington: 1966. 38 pp.

220. NATIONAL AERONAUTICS AND SPACE ADMINISTRATION, SCIENTIFIC AND TECHNICAL INFORMATION DIVISION, *Scientific and Technical Aerospace Reports*, Washington: 1963 to date. Semimonthly.

221. NATIONAL COUNCIL ON MARINE RESOURCES AND ENGINEERING DEVELOPMENT, *Marine Science Activities*, Washington: 1968. 5 vols.

222. NATIONAL SCIENCE FOUNDATION, OFFICE OF SCIENCE INFORMATION SERVICE, *Specialized Science Information Services in the United States—A Directory of Selected Specialized Information Services in the Physical and Biological Sciences*, Washington: 1961. 528 pp.

223. NATIONAL WILDLIFE CONFEDERATION, *Conservation Directory*, Washington: 1967. Annual listing of agencies and organizations.

224. NUPEN, WILHELM, and MALCOLM RIGBY, "Annotated Bibliography on Tropical Cyclones, Hurricanes and Typhoons," *Meteorological Abstracts*, 7 (1956), 1115–1163.

225. *Oceanic Citation Journal*, Oceanic Information Center, P.O. Box 2369, La Jolla, Calif. 92037.

226. *Oceanic Index* (Retrieval Service), Oceanic Information Center, P.O. Box 2369, La Jolla, Calif. 92037.

227. PEARL, RICHARD M., *Guide to Geologic Literature*, New York: McGraw-Hill, 1951. 239 pp.

228. PELLE, WILLIAM J., *Annotated Bibliography on the Planning Aspects of Air Pollution Control*, Northeastern Illinois Planning Commission, 1965. 47 pp.

229. PENNSYLVANIA STATE UNIVERSITY, CENTER FOR AIR ENVIRONMENT STUDIES, *Index to Air Pollution Research*, University Park: 1965 to date. Annual.

230. PRICE, RAYMOND (comp.), *Watershed Management Research in the Southwest*, Ft. Collins, Colo.: Rocky Mountain Forest and Range Experiment Station, 1958. 33 pp.

231. PROUDFOOT, BRUCE, *Current Research in Geomorphology,* Durham, England: Science Laboratories, Durham.

232. RELPH, E. C., and S. B. GOODWILLIE, *Annotated Bibliography on Snow and Ice Problems,* Toronto: University of Toronto, 1968. 14 pp.

233. RIGBY, MALCOLM, "Bibliography of Outstanding Reference Works of the Past Decade," *Meteorological and Geoastrophysical Abstracts,* 11 (January, 1960), 101–143.

234. ROBINSON, T. W., and A. I. JOHNSON, *Selected Bibliography on Evaporation and Transpiration,* U.S. Geological Survey, Water Supply Paper 1539 R, Washington: 1961.

235. RODIER, J., *Bibliography of African Hydrology,* New York: UNESCO, 1963. 166 pp.

236. *Scientific Information Exchange,* Washington: Smithsonian Institution, n.d. (Current research in physical sciences on computer tape.)

237. TOMPKIN, J. M., and S. H. BRITT, *Landslides: A Selected Annotated Bibliography,* Washington: Highway Research Board, 1951. (Bibliography-10.)

238. UNITED NATIONS EDUCATIONAL, SCIENTIFIC AND CULTURAL ORGANIZATION, *World Guide to Science Information and Documentation Services,* New York: 1965. 211 pp.

239. ——, *World Guide to Technical Information and Documentation Services,* New York: 1966.

240. UNITED NATIONS, FOOD AND AGRICULTURE ORGANIZATION, *Index 1954–1966: Land and Water,* Rome: 1968.

241. U.S. DEPARTMENT OF AGRICULTURE, AGRICULTURAL RESEARCH SERVICE, *Abstracts of Recent Published Materials on Soil and Water Conservation,* Washington. Issued irregularly. (Discontinued 1967.)

242. U.S. DEPARTMENT OF AGRICULTURE, FOREST SERVICE, SOUTHEAST FOREST EXPERIMENT STATION, *Fire: A Summary of Literature in the United States from the Mid-1920's to 1966,* Asheville, N.C.: 1968. 117 pp.

243. U.S. DEPARTMENT OF AGRICULTURE, FOREST SERVICE, *Watershed Management,* Technical Paper No. 53, Berkeley: 1960. 15 pp.

244. U.S. DEPARTMENT OF AGRICULTURE, LIBRARY, *Bibliography of Agriculture*, Washington: 1942. Monthly, January 1966 to date.

245. U.S. DEPARTMENT OF AGRICULTURE, SOIL CONSERVATION SERVICE, *Bibliography on Soil and Water Conservation*, Miscellaneous Publication 312, Washington: 1938.

246. U.S. DEPARTMENT OF COMMERCE, ESSA, WEATHER BUREAU, *A Bibliography of Weather and Architecture*, John F. and Joan M. Griffiths (comps.), Silver Spring, Md.: 1969.

247. U.S. DEPARTMENT OF COMMERCE, NATIONAL BUREAU OF STANDARDS, *Hydraulic Research in the United States*, Miscellaneous Publication 205, Middleton and Matchett (eds.), Washington: 1952. 200 pp.

248. U.S. DEPARTMENT OF COMMERCE, WEATHER BUREAU, OFFICE OF HYDROLOGY, *Annotated Bibliography of ESSA Publications of Hydrometeorological Interest*, 2nd ed., Washington: 1967. 27 pp.

249. U.S. DEPARTMENT OF DEFENSE, AIR WEATHER SERVICE, *Catalogue of Local Forecast Studies*, Scott Air Force Base, Ill.: 1961. 135 pp.

250. U.S. DEPARTMENT OF HEALTH, EDUCATION AND WELFARE, *Air Pollution Publications; A Selected Bibliography*, 1963–1966. Washington: 1966. 144 pp.

251. U.S. DEPARTMENT OF THE INTERIOR, BUREAU OF RECLAMATION, *Annotated Bibliography on Hydrology, 1951–54, and Sedimentation, 1950–54, United States and Canada*, Washington: 1956. 207 pp. *Supplement 1955–58, Supplement 1959–62.* 1964, 323 pp.

252. U.S. DEPARTMENT OF THE INTERIOR, BUREAU OF SPORT FISHERIES, *Wildlife Review*, Washington: 1955 to date. Monthly.

253. U.S. DEPARTMENT OF THE INTERIOR, GEOLOGICAL SURVEY, *Abstracts of North American Geology*, Washington. Monthly since January, 1966.

254. ———, *Annotated Bibliography on Hydrology and Sedimentation United States and Canada 1955–58*, Water Supply Paper 1546, H. C. Griggs (comp.), Washington: 1962. 236 pp.

255. ———, *Bibliography and Index of Publications Relating to Ground Water Prepared by the Geological Survey and Cooperating Agencies*, Water Supply Paper 992, Washington: 1947. 412 pp.

256. ———, *Bibliography of Hydrology of the United States 1963*, Water Supply Paper 1863, Washington: 1966. 166 pp.

257. ———, *Bibliography of North American Geology*, Washington: 1887. Annual.

258. ———, *Bibliography of Publications Relating to Ground Water Prepared by the Geological Survey and Cooperating Agencies 1946–55*, Water Supply Paper 1492, Washington: 1957. 203 pp.

259. ———, *Geological and Water Supply Reports and Maps—California*. Published annually and available from Publication Sales Offices. Various paginations.

260. ———, *Geophysical Abstracts*, Washington: 1929. Monthly. Some numbers published by U.S. Bureau of Mines.

261. ———, *State Water-Rights Laws and Related Subjects, A Bibliography*, Washington: 1962.

262. ———, *Status of Reports on Hydrologic Investigations in Nevada*, Carson City: 1966. 10 pp.

263. U.S. DEPARTMENT OF THE INTERIOR, OFFICE OF SALINE WATER, *Conversion of Saline Water: A Bibliography of Saline Water*, Research and Development Progress Report No. 82, Karel A. Kase, Washington: 1963. 59 pp. Also later editions.

264. ———, *Bibliography of Saline Water Conversion Literature*, Washington: 1965. 381 pp.

265. U.S. LIBRARY OF CONGRESS, NATIONAL REFERRAL CENTER FOR SCIENCE AND TECHNOLOGY, A *Directory of Information Resources in the United States: Physical Sciences, Biological Sciences, Engineering*, Washington: 1965. 352 pp.

266. ———, *A Directory of Information Resources in the United States: Water*, Washington: 1966. 248 pp.

267. U.S. LIBRARY OF CONGRESS, SCIENCE AND TECHNOLOGY DIVISION, *Bibliography on Snow, Ice, and Permafrost with Abstracts*, Washington: 1951 to date. (Semiannual 1951–1956; annual 1956 to date.)

268. ———, *Guide to U.S. Indexing and Abstracting Services in Science and Technology*, Washington: 1960.

269. ———, *Guide to the World's Abstracting and Indexing Services in Science and Technology*, National Federation of Science Ab-

stracting and Indexing Services Report No. 102, Washington: 1963. 183 pp.

270. ———, *Weather Modification in the Soviet Union, 1946–1966; A Selected Annotated Bibliography*, N. T. Zikeev and G. A. Donmain, Washington: 1967. 78 pp.

271. ———, *United States IGY Bibliography 1953–1960*, Washington: 1963. 391 pp.

272. U.S. Public Health Service, Environmental Health Service, NAPCA *Abstract Bulletin*, Raleigh, N.C.: National Air Pollution Control Administration, 1969. Quarterly.

273. Wang, Jen Yu, and Gerald Barger, *Bibliography of Agricultural Meteorology*, Madison: University of Wisconsin Press, 1962.

274. Ward, Richard de C., "A Short Bibliography of United States Climatology," *Journal of Geology* (1918), 137–144.

275. *Water Conservation Bibliography*, Bebington, Cheshire, Great Britain: Price's Ltd. Technical Publication No. 7, 1960. 70 pp.

276. *Water Resources Research Catalog*, Washington: 1966 to date. Annual.

277. Wellisch, H. (comp.), *Water Resources Development, 1950–1965; An International Bibliography*, Jerusalem, Israel, Program for Scientific Translations, New York: Daniel Davey, 1967. 135 pp.

278. World Meteorological Organization, *WMO Bibliography on Climatic Fluctuations*, Geneva: 1961. 90 pp.

C. Social Sciences and Humanities (Including Agriculture)

279. American Economic Association, *Index of Economic Journals, 1886–1959*, Homewood, Ill.: Irwin, 1961–1962. 5 vols.

280. The American Historical Association, *Guide to Historical Literature*, George F. Howe et al (eds.), New York: Macmillan Co., 1961. 962 pp.

281. Aronoff, Leah, *Current Information Sources for Community*

Planning Periodicals and Serials, Monticello, Ill.: Council of Planning Librarians, 1967. 68 pp.

282. BAERWALD, DIANE A., *Survey of Planning Information in Standard Reference Books,* Monticello, Ill.: Council of Planning Librarians, 1968. 34 pp.

283. BELL, GWENDOLYN, B. HYMA, and H. O'BRADY, *Annotated Bibliography of the Patterns and Dynamics of Rural Settlements,* Pittsburgh: University of Pittsburgh, 1968. 70 pp.

284. BESTOR, GEORGE C., and HOLWAY R. JONES, *City Planning: A Basic Bibliography of Sources and Trends,* Sacramento: California Council of Civil Engineers and Land Surveyors, 1966. 195 pp.

285. *Biennial Review of Anthropology,* Stanford: Stanford University Press, 1959 to date.

286. BLANCHARD, JOY R., and HAROLD OSTVOLD, *Literature of Agricultural Research,* Berkeley: University of California Press, 1958. 231 pp.

287. BOLAN, LEWIS, *The Role of Urban Planning in the Residential Integration of Middle Class Negroes and Whites,* Monticello, Ill.: Council of Planning Librarians, 1968. 6 pp.

288. *Business Periodicals Index,* New York: H. W. Wilson, Jan. 1958–Jan. 1967. Monthly except July.

289. CARVAJAL, JOAN, and MARTHA E. MUNZER, *Conservation Education: A Selected Bibliography,* Danville, Ill.: Interstate Printers, 1968. 98 pp.

290. CATANESE, ANTHONY JAMES, *Systematic Planning: An Annotated Bibliography and Literature Guide,* Monticello, Ill.: Council of Planning Librarians, 1969. (Exchange Bibliography, 91.)

291. CHAPIN, F. STUART, JR., *Selected References on Urban Planning and Techniques,* Durham, N.C.: University of North Carolina, 1966. 68 pp.

292. CLARK, ROBERT A., *Data Bank or Information Systems Publications with Emphasis on Land Use,* Monticello, Ill.: Council of Planning Librarians, 1968. 10 pp.

293. ———, *Selected References on Land Use Inventory Methods,* Monticello, Ill.: Council of Planning Librarians, 1969. (Exchange Bibliography, 92.)

294. COMAN, EDWIN T., *Sources of Business Information*, 2nd ed., Englewood Cliffs, N.J.: Prentice-Hall, 1964. 406 pp.

295. COULTER, EDITH M., and MELANIE GERSTENFELD, *Historical Bibliographies: A Systematic and Annotated Guide*, Berkeley: University of California Press, 1935. 206 pp. Reissued, New York: Russell and Russell, 1965. 206 pp.

296. COX, EDWARD G., *A Reference Guide to the Literature of Travel*, Seattle: University of Washington. Publications in Language and Literature, Vols. 9, 10, and 12, 1935–49. 3 vols.

297. DENMAN, DONALD R., et al, *Bibliography of Rural Land Economy and Land Ownership, 1900–1957*, Cambridge, England: Cambridge University, Department of Estate Management, 1958. 412 pp.

298. DOBIE, J. FRANK, *Guide to Life and Literature of the Southwest*, Dallas: Southern Methodist University Press, 1952. 222 pp.

299. DYCKMAN, JOHN W., *An Individual Review of Current Planning Literature*, Monticello, Ill.: Council of Planning Librarians, 1967. 18 pp.

300. EASTMAN, ALAN, *State Planning in the Sixties: A Bibliography*, Monticello, Ill.: Council of Planning Librarians, 1970. 8 pp.

301. *Economic Abstracts*, The Hague: Martinus Nijhoff. Semimonthly since 1953.

302. *The Education Index*, New York: H. W. Wilson. Monthly with annual cumulation (indexes the *Journal of Geography*).

303. ELDRIDGE, HOPE T., *The Materials of Demography; A Selected and Annotated Bibliography*, New York: Columbia University Press, 1959. 222 pp.

304. FEDERAL HOUSING ADMINISTRATOR, HOUSING AND HOME FINANCE AGENCY, OFFICE OF THE ADMINISTRATOR, *Housing and Planning References*, Washington. Bimonthly.

305. FOLEY, DONALD L., *The Metropolitan Region: A Selective Bibliography*, Monticello, Ill.: Council of Planning Librarians, 1968. 9 pp.

306. *Foreign Affairs Bibliography*, New York: Council on Foreign Relations, 1933–1964. (10-year intervals.)

307. GOELDNER, C. R., *Bibliography of Tourism and Travel Research*

Studies, Reports, and Articles, Boulder: University of Colorado, Graduate School of Business, 1967. 70 pp.

308. GOODMAN, WILLIAM I., *Planning Legislation and Administration, An Annotated Bibliography,* Monticello, Ill.: Council of Planning Librarians, 1968. 13 pp.

309. HAZELWOOD, ARTHUR, *The Economics of "Under-developed" Areas; An Annotated Reading List of Books, Articles and Official Publications,* 2nd ed., London: Oxford University Press, 1959. 156 pp. (1st ed. 1954.)

310. ——, *The Economics of Development; An Annotated List of Books and Articles Published 1958–1962,* London: Oxford University Press, 1964. 104 pp.

311. *Historical Abstracts,* Santa Barbara: American Bibliographical Center. 1955 to date. Quarterly.

312. HOWARD, WILLIAM A., *Concept of an Optimum Size City,* A *Selected Bibliography,* Monticello, Ill.: Council of Planning Librarians, 1968. 5 pp.

313. ——, *Geographic Aspects of Urban Planning,* A *Selected Bibliography,* Monticello, Ill.: Council of Planning Librarians, 1968. 8 pp.

314. ——, *Remote Sensing of the Urban Environment,* Monticello, Ill.: Council of Planning Librarians, 1969.

315. INGERSOLL, PHYLLIS W., *Ideal Forms of Cities: An Historical Bibliography,* Monticello, Ill.: Council of Planning Librarians, 1959. 67 pp.

316. INTERNATIONAL ASSOCIATION OF AGRICULTURAL LIBRARIES AND DOCUMENTALISTS, *World Agricultural Economics and Rural Sociology Abstracts,* Amsterdam: 1959 to date. Quarterly.

317. *International Bibliography of Historical Sciences,* New York: H. W. Wilson, 1926. Annual.

318. INTERNATIONAL COMMITTEE FOR SOCIAL SCIENCES DOCUMENTATION, *The International Bibliography of the Social Sciences,* London: Tavistock Pub. Ltd. Annual.

319. *The Journey to Work, Selected References, 1960–1967,* Monticello, Ill.: Council of Planning Librarians, 1968. 8 pp.

320. KAPLAN, LOUIS (ed.), *Research Materials in the Social Sciences,*

rev. ed., Jack A. Clarke, Madison: University of Wisconsin Press, 1959. 42 pp.

321. KARL, KENYON F., *Industrial Parks and Districts: An Annotated Bibliography*, Monticello, Ill.: Council of Planning Librarians, 1968. 14 pp.

322. KUEHL, WARREN F., *Dissertations in History; An Index to Dissertations Completed in History Departments of United States and Canadian Universities, 1873–1960*, Lexington: University of Kentucky, 1965. 264 pp.

323. LARIMORE, ANN E. (comp.), *World Urbanization and Urban Migration: An Annotated Bibliography*, Ann Arbor: Department of Geography, University of Michigan, 1969. 148 pp.

324. LEWIS, PETER R., *The Literature of the Social Sciences; An Introductory Survey and Guide*, London: Library Association, 1960. 222 pp.

325. MACKESEY, THOMAS W., *History of City Planning*, Monticello Ill.: Council of Planning Librarians, 1961. 58 pp.

326. MANGALAM, J. J., *Human Migration: A Guide to Migrant Literature in English, 1955–1962*, Lexington: University of Kentucky Press, 1968. 194 pp.

327. MANLEY, M. C., *Business Information: How to Find and Use It*, New York: Harper, 1955. 265 pp.

328. MENGES, GARY L., *Model Cities*, Monticello, Ill.: Council of Planning Librarians, 1968. 13 pp.

329. METCALF, KENNETH N., *Transportation: Information Sources*, Detroit: Gale Research, 1965. 307 pp.

330. MUKHERJEE, AJIT K., *Annotated Guide to Reference Materials in the Human Sciences*, Bombay and New York: Asia Publishing House, 1963. 267 pp.

331. NEVILLE, ROLFE E. (comp.), *Economic Aspects of Agricultural Development in Africa*, Oxford: University of Oxford Agricultural Economics Research Institute, 1969. Unpaged.

332. NORTHWESTERN UNIVERSITY, TRANSPORTATION CENTER LIBRARY, *Current Literature in Traffic and Transportation*, Evanston: 1960 to date. Monthly.

333. ———, *A Reference Guide to Metropolitan Transportation*, Evanston: 1964. 42 pp.

334. ———, *Sources of Information in Transportation*, Evanston: 1964. 262 pp.

335. OLSSON, GUNNAR, *Distance and Human Interaction: A Review and Bibliography*, Bibliography Series No. 2, Philadelphia: Regional Science Research Institute, 1965.

336. PAN AMERICAN UNION, *The Plantations: A Bibliography*, Edgar T. Thompson (ed.), Washington: 1957.

337. PILLAI, N. G., and JOYCE LING, *Regional Development and Economic Growth: The Problem Background, A Select Bibliography*, Monticello, Ill.: Council of Planning Librarians, 1970. 19 pp.

338. PILLAI, N. G. and others, *Regional Development and Economic Growth: Theory, Analysis and Techniques: A Select Bibliography*, Monticello, Ill.: Council of Planning Librarians, 1970. 18 pp.

339. PINKERTON, JAMES R. and MARJORIE J. PINKERTON, *Outdoor Recreation and Leisure: A Reference Guide and Selected Bibliography*, Columbia: University of Missouri, 1969. 332 pp.

340. POLLACK, LESLIE J., *Driver Distraction as Related to Physical Development Abutting Urban Streets: An Empirical Inquiry into the Design of the Motorists Visual Environment*, Monticello, Ill.: Council of Planning Librarians, 1968. 4 pp.

341. PRINCETON UNIVERSITY, OFFICE OF POPULATION RESEARCH AND POPULATION ASSOCIATION OF AMERICA, *Population Index*, Princeton: 1935–July, 1966. Quarterly.

342. *Public Affairs Information Service Bulletin*, New York: Public Affairs Information Service. Weekly.

343. *Research Digest*, University of Illinois, Bureau of Community Planning, Vol. 1, 1954. 2 months a year plus annual index.

344. RICKERT, JOHN, with the collaboration of Jerome P. Pickard, *Open Space Land Planning and Taxation: A Selected Bibliography*, Washington: Government Printing Office, 1965.

345. ROBBINS, JANE B., *Access to Airports: Selected References*, Monticello, Ill.: Council of Planning Librarians, 1968. 21 pp.

346. ROYAL ANTHROPOLOGICAL INSTITUTE OF GREAT BRITAIN AND IRELAND LIBRARY, *Index to Current Periodicals Received in the Library of the Royal Anthropological Institute*, Vol. 1, 1963. Quarterly.

347. ROYAL INSTITUTE OF INTERNATIONAL AFFAIRS, LIBRARY, LONDON,

Index to Periodical Articles, 1950–1964, Boston: G. K. Hall, 1964. 2 vols.

348. SAN FERNANDO VALLEY STATE COLLEGE LIBRARY, *Black Brown Bibliography*, Northridge, Calif.: 1968. 86 pp.

349. SCHNEIDERMEYER, MELVIN, *The Metropolitan Social Inventory: Procedures for Measuring Human Well-Being in Urban Areas*, Monticello, Ill.: Council of Planning Librarians, 1968. 8 pp.

350. SCOTT, FRANKLIN D., and ELAINE TEIGLER, *Guide to the American Historical Review, 1895–1945: A Subject-Classified Explanatory Bibliography of the Articles, Notes and Suggestions, and Documents*, The American Historical Association Annual Report for the year 1944, Vol. I, Part 2; Washington: Government Printing Office, 1945. pp. 65–292.

351. SHILLABER, CAROLINE, *References on City and Regional Planning*, Massachusetts Institute of Technology, Technology Monographs, Library Series No. 2, Cambridge: Technology Press, 1960. 41 pp.

352. *Sociological Abstracts*, New York: 1953. Quarterly.

353. SPIELVOGEL, SAMUEL, *A Selected Bibliography on City and Regional Planning*, Washington: Scarecrow Press, 1951.

354. STEVENS, BENJAMIN H., and CAROLYN A. BRACKETT, *Industrial Location; A Review and Annotated Bibliography of Theoretical, Empirical and Case Studies*, Philadelphia: Regional Science Research Institute, 1967. 199 pp.

355. STOOTS, F., *Regional Planning, An Introductory Bibliography*, Monticello, Ill.: Council of Planning Librarians, 1968. 5 pp.

356. SUTTLES, PATRICIA H. (comp.), *Educators Guide to Free Social Studies Materials*, Randolph, Wis.: Educators Progress Service, 1966. 480 pp.

357. TEXAS, UNIVERSITY OF, POPULATION RESEARCH CENTER, *International Population Census Bibliography*, Austin: various dates.

358. TUCKER, DOROTHY, *Computers and Information Systems in Planning and Related Government Functions*, Monticello, Ill.: Council of Planning Librarians, 1968. 21 pp.

359. UNITED NATIONS, ECONOMIC AND SOCIAL COUNCIL, *Catalogue of Economic and Social Projects*, New York: 1946.

360. UNITED NATIONS EDUCATIONAL, SCIENTIFIC, AND CULTURAL OR-
GANIZATION, *International Bibliography of Economics,* Paris:
1952. Annual.

361. ———, *International Bibliography of Political Science,* Paris:
1953. Annual.

362. ———, *International Bibliography of Social and Cultural Anthropology,* Paris: 1955. Annual.

363. ———, *International Bibliography of Sociology,* Paris: 1952.
Annual.

364. ———, *International Political Science Abstracts,* Paris: 1951.
Quarterly.

365. ———, *World List of Social Science Periodicals,* 2nd ed., Paris:
1957. 210 pp.

366. UNITED NATIONS, FOOD AND AGRICULTURE ORGANIZATION, *Bibliography on Land Tenure,* Rome: 1955. 386 pp. Supplement, Rome:
1959.

367. ———, *World Fisheries Abstracts,* Rome: 1950 to date. Quarterly.

368. U.S. DEPARTMENT OF AGRICULTURE, *Urbanization and Changing
Land Uses: A Bibliography of Selected References,* 1950–1958,
Elizabeth G. Davis, Hugh A. Johnson, and Claude C. Haren,
Miscellaneous Publication 825, Washington: 1960.

369. U.S. DEPARTMENT OF AGRICULTURE, ECONOMIC RESEARCH SERVICE,
Research Data on Minority Groups (1955–1965), Miscellaneous
Publication 1046, Washington: 1966. 25 pp.

370. U.S. DEPARTMENT OF THE INTERIOR, OFFICE OF WATER RE-
SOURCES RESEARCH, *Bibliography on Socio-Economic Aspects of
Water Resources,* H. R. Hamilton, et al, Washington: 1966.
453 pp.

371. U.S. LIBRARY OF CONGRESS, NATIONAL REFERRAL CENTER FOR
SCIENCE AND TECHNOLOGY, *A Directory of Information Resources
in the United States: Social Sciences,* Washington: 1965. 218 pp.

372. U.S. NATIONAL AGRICULTURAL LIBRARY, *Bibliography of Agriculture,* Washington: 1942 to date. Monthly.

373. U.S. NATIONAL HISTORICAL PUBLICATIONS COMMISSION, *Guide to*

Archives and Manuscripts in the United States, Philip M. Hamer (ed.), New Haven: Yale University Press, 1961. 775 pp.

374. WHEATON, WILLIAM L. C., WILLIAM C. BAER, and DAVID M. VETTER, *Housing, Renewal, and Development Bibliography,* Monticello, Ill.: Council of Planning Librarians, 1968. 94 pp.

375. WHEELER, JAMES O., *Research on the Journey to Work: Introduction and Bibliography,* Monticello, Ill.: Council of Planning Librarians, 1969. 21 pp.

376. VANCE, MARY, *State Outdoor Recreation Plans,* Monticello, Ill.: Council of Planning Librarians, 1967. 7 pp.

377. WILCOXEN, RALPH, *A Short Bibliography on Megastructures,* Monticello, Ill.: Council of Planning Librarians, 1969. 18 pp.

378. WISCONSIN UNIVERSITY LAND TENURE CENTER, LIBRARY, *Agrarian Reform and Land Tenure: A List of Source Materials, with Special Sections on Agricultural Finance, Taxation and Agriculture, Agricultural Statistics, Bibliographical Sources,* Madison: 1965. 105 pp.

D. Regional—United States

379. ADAMS, RAMON F., *Burrs Under the Saddle: A Second Look at Books and Histories of the West,* Norman: University of Oklahoma Press, 1964. 610 pp.

380. *America: History and Life, A Guide to Periodical Literature,* Santa Barbara: American Bibliographical Center. Quarterly with No. 4 the annual index.

381. AMERICAN ASSOCIATION FOR STATE AND LOCAL HISTORY, *Directory of Historical Societies and Agencies, in the United States and Canada,* Columbus, Ohio: 1956 to date.

382. BEATTY, W. B., *Mineral Resource Data in the Western States,* Stanford: Stanford University School of Mineral Sciences, 1962. 42 pp. (Sources of information and maps.)

383. BEERS, HENRY P., *Bibliographies in American History: Guide to Materials for Research,* New York: H. W. Wilson, 1942. 487 pp. Reprinted, Paterson, N.J.: Pageant Books, 1959. 487 pp.

384. BERRY, BRIAN J. L., and THOMAS D. HANKINS, *A Bibliographic Guide to the Economic Regions of the United States*, Chicago: University of Chicago, Department of Geography, 1963. 101 pp.

385. BRUGGE, DAVID M., J. LEE CORREL, and EDITHA L. WATSON, *Navajo Bibliography*, Window Rock, Ariz.: The Navajo Tribe, 1967. 291 pp.

386. CALIFORNIA LIBRARY ASSOCIATION, *California Local History*, Stanford: Stanford University Press, 1950. 576 pp.

387. CALIFORNIA, UNIVERSITY OF, BUREAU OF PUBLIC ADMINISTRATION, *Land Utilization: A Bibliography*, Berkeley: 1935. 222 pp. Supplement, 1937. 139 pp.

388. CARNEGIE INSTITUTION OF WASHINGTON, *Index of Economic Material in Documents of the States of the United States: California, 1849–1904*, Adelaide R. Hasse, Washington: 1908. 316 pp.

389. CHAMBER OF COMMERCE OF THE STATE OF NEW YORK, *Directory of Chambers of Commerce in the United States*, New York: 1922 to date. Annual listing of chambers of cities of 5,000 and over by state.

390. COAN, O. W., and R. G. LILLARD, *America in Fiction: An Annotated List of Novels That Interpret Aspects of Life in the United States*, Palo Alto: Pacific Books, 1967. 232 pp.

391. CONWAY, H. McKINLEY, *Area Development Organizations*, Atlanta: Conway Research Inc., 1966. 331 pp.

392. THE COUNCIL OF STATE GOVERNMENTS, *The Book of the States*, Chicago. Biennial. Lists state officials and offices.

393. COWAN, ROBERT E. and ROBERT G. COWAN, *A Bibliography of the History of California, 1510–1930*, San Francisco: John Henry Nash, 1933. 3 vols. Reprinted 1964 by Dawson's Book Shop.

394. COWAN, ROBERT E., and BOUTWELL DUNLAP, *Bibliography of the Chinese Question in the United States*, San Francisco: A. M. Robertson, 1909. 68 pp.

395. DAY, A. GROVE, *Coronado's Quest, The Discovery of the Southwest States*, Berkeley: University of California Press, 1964. 419 pp.

396. DESKINS, DONALD R. (comp.), *Geographical Literature on the American Negro, 1949–1968, A Bibliography*, Ann Arbor: University of Michigan, 1968. 15 pp.

397. *Directory of the Forest Products Industry*, Portland, Ore.: Miller Freeman. Annual. (Includes statistics.)

398. EDWARDS, E. I., *The Enduring Desert: A Descriptive Bibliography*, Los Angeles: Ward Ritchie Press, 1969.

399. EDWARDS, F., *Bibliography of the History of Agriculture in the United States*, Washington: 1930.

400. *Encyclopedia of American Associations*, Detroit: Gale Research, 1961.

401. EVANS, HENRY H., *Western Bibliographies*, San Francisco: The Peregrine Press, 1951. 40 pp.

402. FARQUHAR, FRANCIS P., *The Books of the Colorado River and the Grand Canyon: A Selective Bibliography*, Los Angeles: Glen Dawson, 1953. 75 pp.

403. GOHDES, CLARENCE, *Bibliographical Guide to the Study of the Literature of the U.S.A.*, 3rd ed., Durham, N.C.: Duke University Press, 1963.

404. HANDLIN, OSCAR, et al, *Harvard Guide to American History*, Cambridge: Harvard University Press, 1954. 689 pp.

405. HASKELL, DANIEL (comp.), *The United States Exploring Expedition, 1838–1842, and Its Publications, 1844–1874; A Bibliography*, New York: The New York Public Library, 1942. 188 pp.

406. McMANIS, DOUGLAS R., *Historical Geography of the United States: A Bibliography*, Ypsilanti, Mich.: Eastern Michigan University, Division of Field Services, 1965. 249 pp.

407. MEISEL, MAX, *A Bibliography of American Natural History: The Pioneer Century, 1769–1865*, Brooklyn: The Premier Publishing Co., 1924–1929. 3 vols.

408. MILLER, ELIZABETH W., *The Negro in America: A Bibliography*, Cambridge: Harvard University Press, 1966.

409. NEEDHAM, PAUL E. (ed.), *Plant Location*, New York: Simmons-Boardman Publishing Corp., 1959 to date. Annual.

410. NEWBERRY LIBRARY, *Dictionary Catalog of the Edward E. Ayer Collection of Americana and the American Indian*, Chicago: 1961. 16 vols.

411. NOLTING, ORIN F., *The Municipal Yearbook*, Chicago: The International City Managers' Association, 1933 to date. (Annual

listing of sources of information and of statistics of American cities.)

412. NORTHEASTERN ILLINOIS PLANNING COMMISSION, *A Selected Bibliography on the Chicago Metropolitan Area*, Chicago: 1964. 112 pp.

413. PAYLORE, PATRICIA, *Seventy-five Years of Arid Lands Research at the University of Arizona; A Selective Bibliography, 1891–1965*, Tucson: Arid-Lands Research, University of Arizona, 1966. 95 pp.

414. PETERSON, CLARENCE S., *Consolidated Bibliography of County Histories in Fifty States in 1961, Consolidated 1935–1961*, 2nd ed., Baltimore: Genealogical Publishing Co., 1963. 186 pp.

415. POWELL, LAWRENCE C., *Books West Southwest*, Los Angeles: Ward Ritchie Press, 1957. 157 pp.

416. ROORBACH, ORVILLE A., *Bibliotheca Americana, 1820–1861*, New York: Peter Smith, 1939. 4 vols.

417. ROSS, FRANK A., and LOUISE V. KENNEDY, *A Bibliography of Negro Migration*, New York: Columbia University Press, 1934.

418. SABIN, JOSEPH, *Bibliotheca Americana, A Dictionary Guide for Books Relating to America, from Its Discovery to the Present Time*, Amsterdam: Bibliographical Society of America, 1961–1962. 29 vols.

419. SEALOCK, RICHARD B., and PAULINE A. SEELY, *Bibliography of Place-Name Literature: United States and Canada*, 2nd ed., Chicago: American Library Association, 1967. 352 pp.

420. SMITHSONIAN INSTITUTION, BUREAU OF AMERICAN ETHNOLOGY, *An Analysis of Sources of Information on the Population of the Navajo*, Bulletin 197 by Denis F. Johnston, Washington: 1966. 220 pp.

421. SNODGRASS, MARJORIE P., *Economic Development of American Indians and Eskimos, 1930 through 1967, A Bibliography*, Washington: Bureau of Indian Affairs, 1968.

422. TENNESSEE VALLEY AUTHORITY, TECHNICAL LIBRARY, *A Bibliography for the TVA Program*, Knoxville: 1964. 67 pp.

423. TURNER, FREDERICK J., *List of References on the History of the West*, Cambridge: Harvard University Press, 1920. 133 pp.

424. U.S. DEPARTMENT OF AGRICULTURE, *Bibliography on Land Settle-*

ment, Miscellaneous Publication 172, Louise O. Bercow and others, Washington: 1934. 492 pp.

425. ———, *Bibliography on Land Utilization, 1918–36*, Miscellaneous Publication 284, Louise O. Bercow and Annie M. Hannay, Washington: 1938. 1508 pp.

426. ———, *The Economics of Irrigation in the United States, A List of Annotated References, 1950–1957*, E. G. Davis, E. J. Suber, and W. F. Ehlers, Washington: 1958. 66 pp.

427. ———, *Land Ownership, A Bibliography of Selected References*, Annie M. Hannay and Donald W. Gooch, Washington: 1953. 293 pp.

428. U.S. DEPARTMENT OF THE INTERIOR, BUREAU OF LAND MANAGEMENT, *Public Lands Bibliography*, Washington: 1962. 106 pp. Supplement, 1965. 71 pp.

429. U.S. LIBRARY OF CONGRESS, *A Guide to the Study of the United States of America; Representative Books Reflecting the Development of American Life and Thought*, Donald H. Muggridge and Blanche P. McCrum, Washington: 1960. 1193 pp.

430. WAGNER, HENRY R., *The Spanish Southwest*, New York: Arno Press, 1967. 2 vols. (Facsimile reproduction of 1937 edition.)

431. WEBER, FRANCIS J. (Reverend), *A Bibliography of California Bibliographies*, Los Angeles: Ward Ritchie Press, 1968.

432. WINTHER, OSCAR O., *A Classified Bibliography of the Periodical Literature of the Trans-Mississippi West, 1811–1957*, Bloomington: Indiana University Press, 1961. 626 pp.

433. ———, *The Trans-Mississippi West: A Guide to Its Periodical Literature, 1811–1938*, Bloomington: Indiana University Press, 1942. 263 pp.

434. WISCONSIN, UNIVERSITY OF, LAND TENURE CENTER, LIBRARY, *Agrarian Reform and Land Tenure*; A List of Source Materials with Special Sections on Agricultural Finance, Taxation, and Agriculture, Agricultural Statistics, Bibliographical Sources, Madison: 1965. 105 pp.

435. YALE UNIVERSITY LIBRARY, *Catalog of the Yale Collection of Western Americana*, New Haven: 1961. 4 vols.

E. Regional—World and Parts of the World Outside the United States

436. *African Abstracts, A Quarterly Review of Ethnological, Social and Linguistic Studies Appearing in Current Periodicals,* 1950.

437. AFRICAN BIBLIOGRAPHIC CENTER, *A Current Bibliography on African Affairs,* Vol. 1, Washington: 1962. Bimonthly with annual cumulation.

438. AFRICAN BIBLIOGRAPHIC CENTER, *Special Bibliography Series,* Vol. 1, Washington: 1963 to date.

439. *Afrika-Schriftum: Bibliographie Deutschsprachiger Wissenschaftlicher Veröffentlichungen Über Afrika Südlich der Sahara,* Wiesbaden: Franz Steiner Verlag, 1966. 688 pp. References on the geography, ethnology, linguistics, tropical medicine, zoology, and botany of Africa in the German language.

440. *The American Bibliography of Russian and East European Studies for* 1957, Bloomington: Indiana University Press. Annual.

441. AMERICAN LIBRARY ASSOCIATION, *Guide to Japanese Reference Books,* Chicago: 1966. 303 pp.

442. AMERICAN UNIVERSITIES FIELD STAFF, INC., *A Select Bibliography; Asia, Africa, Eastern Europe, Latin America,* Philip Talbor (ed.), New York: 1960. 534 pp. Supplements, 1961, 1963.

443. ARCTIC INSTITUTE OF NORTH AMERICA, *Arctic Bibliography,* Washington: Government Printing Office, 1953–1965, Vols. 1–12. Montreal: McGill University Press, 1967, Vol. 13.

444. ASHER, GEORGE MICHAEL, *A Bibliography and Historical Essay on the Dutch Books and Pamphlets Relating to New Netherland and to the Dutch West India Company* . . . , Amsterdam: 1966. 338 pp.

445. *Australian National Bibliography,* Canberra: National Library of Australia. Monthly since 1961 with annual cumulation.

446. BARRETT, ELLEN C., *Baja California, 1535–1964; A Bibliography*

of Historical, Geographical, and Scientific Literature, Los Angeles: Western-lore, 1967. 250 pp. (An earlier edition in 1957.)

447. BARTLETT, HARLEY H., *Fire in Relation to Primitive Agriculture and Grazing in the Tropics; Annotated Bibliography*, Ann Arbor: University of Michigan Botanical Gardens, 1955–57. 3 vols.

448. BARWICK, G. F. (ed.), *The Aslib Directory: A Guide to Sources of Information in Great Britain and Ireland*, London: Oxford University Press, 1928.

449. BATTELLE MEMORIAL INSTITUTE, COLUMBUS, OHIO, *A Guide to the Scientific and Technical Literature of Eastern Europe*, Washington: National Science Foundation, 1963. 94 pp. Prepared for the National Science Foundation.

450. BERTON, PETER and EUGENE W., *Contemporary China: A Research Guide*, Howard Koch, Jr. (ed.), Stanford: Hoover Institution, 1967.

451. BORCHARDT, D. H., *Australian Bibliography; A Guide to Printed Sources of Information*.

452. *Bibliography of Asian Studies*, 1941. Annual. (In *Journal of Asian Studies*, Sept. issue.) (Formerly *Far Eastern Bibliography*, in the *Far Eastern Quarterly*, 1941–1945, annual since 1945. Title varies.)

453. BOSTON UNIVERSITY LIBRARIES, *Catalog of African Government Documents and African Area Index*, Mary D. Herricks (comp.), Boston: G. K. Hall, 1964. 470 pp.

454. BURKETT, J. (ed.), *Special Library and Information Services in the United Kingdom*, London: The Special Library Association, 1961. 200 pp.

455. CANADA, DEPARTMENT OF MINES AND TECHNICAL SURVEYS, GEOGRAPHICAL BRANCH, *Bibliographical Series*, 1, 1950, Ottawa: The Queen's Printer. Irregular.

456. CANADA, DEPARTMENT OF REGIONAL ECONOMIC EXPANSION PLANNING DEVELOPMENT, *Regional Developments and Economic Growth: Problems, Analyses, and Policies; A Select Bibliography*, Ottawa: 1969.

457. *Canadian Almanac and Directory*, Toronto: Copp, Clark Co., 1848 to date. Annual directory.

458. CANADIAN LIBRARY ASSOCIATION, *Canadian Index, A Guide to*

Canadian Periodicals and Documentary Films, Ottawa: 1947 to date. Monthly guide to over 60 Canadian periodicals.

459. CHAMBER OF COMMERCE OF THE UNITED STATES, FOREIGN COMMERCE DEPARTMENT, *Guide to Foreign Information Sources,* rev. ed., Washington: 1957.

460. CHICAGO, UNIVERSITY OF, PHILIPPINE STUDIES PROGRAM, *Selected Bibliography of the Philippines, Topically Arranged and Annotated,* Fred Eggan and others, New Haven, Conn.: Human Relations Area Files. Preliminary ed., 1956. 238 pp.

461. CHILE, UNIVERSITY OF, INSTITUTE OF GEOGRAPHY, *Informaciónes Geograficas,* Santiago: 1964. 2 vols. (1st vol. general articles; 2nd vol. bibliography on Chile and a selected general geographical bibliography.)

462. CRAIG, ALAN K., *Marine Desert Ecology of Southern Peru,* Springfield, Va.: Clearinghouse, Dept. of Commerce, 1968. 215 pp.

463. CRAIG, ALAN K., and NORBERT P. PSUTY, *1000 Selected References on the Geography, Oceanography, Geology, Ecology, and Archeology of Coastal Peru and Adjacent Areas,* Madison: University of Wisconsin, Department of Geography, 1968.

464. *Directory of Selected Research Institutes in Eastern Europe,* New York: Columbia University Press, 1967.

465. DOST, H., *Bibliography on Land and Water Utilization in the Middle East,* Wageningen, Netherlands: 1953. 115 pp.

466. ECONOMIC RESEARCH INSTITUTE, *A Selected and Annotated Bibliography of Economic Literature on the Arabic Speaking Countries of the Middle East: 1938–1952,* Beirut: American University of Beirut, 1954. Annual supplements.

467. EDINBURGH, UNIVERSITY OF, DEPARTMENT OF SOCIAL ANTHROPOLOGY, *African Urbanization: A Reading List of Selected Books, Articles and Reports,* London: International African Institute, 1965. 27 pp.

468. EMBREE, AINSLIE T., et al (comps.), *Asia, A Guide to Basic Books,* New York: The Asia Society, 1966. 57 pp.

469. EMBREE, JOHN F., and LILLIAN O. DOTSON, *Bibliography of the Peoples and Cultures of Mainland Southeast Asia,* New Haven: Yale University, Southeast Asia Studies, 1950. 821 pp.

470. ETTINGHAUSEN, RICHARD, *A Selected and Annotated Bibliography of Books and Periodicals in Western Languages Dealing with the Near and Middle East; With Special Emphasis on Mediaeval and Modern Times*, Washington: Middle East Institute, 1954. 137 pp.

471. *The Europa Year Book*, London: Europa Publications Ltd., 1964. 2 vols. (Statistics plus a listing of private and governmental organizations in all parts of the world.)

472. FERGUSON, JOHN A., *Bibliography of Australia*, Sydney: Angus and Robertson, 1941 to date. 7 vols.

473. FIELD, HENRY, *Bibliography on Southwestern Asia*, Coral Gables, Fla.: University of Miami Press, 1954 to date. 8 vols.

474. FLYNN, ALICE H., *World Understanding; A Selected Bibliography*, Dobbs Ferry, N.Y.: Oceana Publications, 1965. 263 pp.

475. GOLANY, GIDEON, *City and Regional Planning and Development in Israel*, Monticello, Ill.: Council of Planning Librarians, 1968. 30 pp.

476. ———, *National and Regional Planning and Development in the Netherlands, An Annotated Bibliography*, Monticello, Ill.: Council of Planning Librarians, 1968. 38 pp.

477. ———, *Regional Planning and Development in Developing Countries with Emphasis on Asia and the Middle East*, Monticello, Ill.: Council of Planning Librarians, 1968. 15 pp.

478. GROPP, ARTHUR E. (comp.), *A Bibliography of Latin American Bibliographies*, Metuchen, N.J.: Scarecrow Press, 1968. 515 pp.

479. GUZMAN, LOUIS E., *An Annotated Bibliography of Publications on Urban Latin America*, Chicago: University of Chicago, Department of Geography, 1953. 53 pp.

480. HALL, ROBERT B., and TOSHIO NOH, *Japanese Geography: A Guide to Japanese Reference and Research Materials*, Center for Japanese Studies, Bibliographical Series No. 6, Ann Arbor: University of Michigan Press, 1956. 128 pp.

481. HERMAN, THEODORE, *The Geography of China; A Selected and Annotated Bibliography*, Publication No. 7 of the State Education Department, Albany: University of the State of New York, 1967.

482. HILLS, THEO. L. (comp.), *A Directory of Institutions Primarily*

Devoted to Humid Tropics Research (International Geographical Union, Humid Tropics Commission), Montreal: Department of Geography, McGill University, 1965. 222 pp.

483. HISPANIC and LUSO-BRAZILIAN COUNCILS, *Latin America; An Introduction to Modern Books in England Concerning the Countries of Latin America*, 2nd ed., rev., London: Library Association, 1966. 41 pp.

484. HORECKY, PAUL L. (ed.), *Basic Russian Publications: An Annotated Bibliography on Russia and the Soviet Union*, Chicago: University of Chicago Press, 1962. 313 pp.

485. ———, *Russia and the Soviet Union; A Bibliographic Guide to Western Language Publications*, Chicago: University of Chicago Press, 1965. 478 pp.

486. HUCKER, CHARLES O., *China: A Critical Bibliography*, Tucson: University of Arizona Press, 1962. 136 pp.

487. HUKE, ROBERT E., *Bibliography of Philippine Geography, 1940–1963*, Hanover, N.H.: Dartmouth, Department of Geography, 1964. 84 pp.

488. *Index to Latin American Periodicals, Humanities and Social Sciences*, New York: Scarecrow Press, Vol. 1 (1961). Quarterly, with annual cumulations. Prepared by the Columbus Memorial Library of the Pan American Union and the New York Public Library. Volumes for 1961–62 published in Boston by G. K. Hall.

489. *Indice General de Publicaciónes Periódicas Latinoamericanas; humanidades y ciencias sociales*, 1961 to date. Quarterly with annual cumulations.

490. JUNOD, VIOLAINE I., and IDRIAN N. RESNICK, *The Handbook of Africa*, New York: New York University Press, 1963. 472 pp.

491. KENNEDY, RAYMOND, *Bibliography of Indonesian Peoples and Cultures*, 2nd ed., New Haven: Yale University, Southeast Asia Studies, 1962. 207 pp.

492. *Latin America, An Annotated Bibliography of Paperback Books*, Washington: 1967. 77 pp.

493. LOGAN, MARGUERITE, *Geographical Bibliography for all the Major Nations of the World; Selected Books and Magazine Articles*, Ann Arbor: 1959. 396 pp.

494. LYSTAD, ROBERT A. (ed.), *The African World: A Survey of Social*

Research, edited for the African Studies Association, New York: Frederick A. Praeger, 1965.

495. MACRO, ERIC, *Bibliography of the Arabian Peninsula,* Coral Gables, Fla.: University of Miami, 1958.

496. MAHAR, J. MICHAEL, *India: A Critical Bibliography,* Tucson: University of Arizona Press, 1964.

497. MAICHEL, KAROL, *Guide to Russian Reference Books,* Stanford: Hoover Institution, Bibliographical Series X, 1962 to date. 6 vols.

498. NEAL, J. A. (comp.), *Reference Guide for Travelers,* New York: R. R. Bowker, 1969. 674 pp.

499. NEW YORK PUBLIC LIBRARY, *Dictionary Catalogue of the Schomburg Collection of Negro Literature and History,* New York: 1962. 9 vols.

500. THE NEW YORK PUBLIC LIBRARY, REFERENCE DEPARTMENT, *Dictionary Catalogue of the History of the Americas Collection,* Boston: G. K. Hall, 1961. 28 vols.

501. ———, *Dictionary Catalog of the Oriental Collection,* Boston: G. K. Hall, 1960. 16 vols.

502. NEW YORK UNIVERSITY, BURMA RESEARCH PROJECT, *Annotated Bibliography of Burma,* Frank N. Trager, John N. Musgrave, and Janet Welsh, New Haven: Human Relations Area Files, 1956. 230 pp.

503. NORTHWESTERN UNIVERSITY LIBRARY, *Catalog of the African Collection,* Evanston, Ill.: 1962. 2 vols.

504. O'LEARY, TIMOTHY J., *Ethnographic Bibliography of South America,* New Haven: Human Relations Area Files, 1963. 387 pp.

505. OLSON, RALPH E., *The Literature of Regional Geography: A Check List for University and College Libraries,* National Council for Geographic Education, Special Publication No. 5, Norman, Okla.: 1960. 19 pp.

506. PAN AMERICAN UNION, *Guide to Latin American Scientific and Technical Periodicals; An Annotated List,* Washington: 1962.

507. ———, *Inter-American Committee for Agricultural Development (CIDA); Inventory of Information Basic to the Planning of Agricultural Development in Latin America; Selected Bibliography,* Washington: Organization of American States, 1964. 187 pp.

508.   ———, *Inter-American Review of Bibliography*, Washington: Quarterly since 1951.

509.   PAN AMERICAN UNION, COLUMBUS MEMORIAL LIBRARY, *Index to Latin American Periodical Literature, 1929–1960*, Boston: G. K. Hall, 1962. 8 vols. 1961–1965, 1967, 2 vols.

510.   PATAI, RAPHAEL, *Jordan, Lebanon and Syria: An Annotated Bibliography*, New Haven: Human Relations Area Files, 1957.

511.   PATTERSON, MAUREEN L. P., and INDEN, RONALD B. (eds.), *Introduction to the Civilization of India; South Asia, An Introductory Bibliography*, Chicago: University of Chicago, The College, 1962. 412 pp.

512.   PAYLORE, PATRICIA, *Arid Lands Research Institutions: A World Directory*, Tucson: University of Arizona Press, 1967. 268 pp.

513.   PECSI, MORTON, *Ten Years of Physicogeographic Research in Hungary*, Budapest: Hungarian Academy of Sciences Geographical Research Institute, 1964. 131 pp.

514.   PELZER, KARL J., *Selected Bibliography on the Geography of Southeast Asia*, New Haven: Yale University, Southeast Asian Studies, 1949. 3 vols.

515.   RUGGLES, MELVILLE J., and VACLAV MOSTECKY, *Russian and East European Publications in the Libraries of the United States*, New York: Columbia University Press, 1960. 396 pp.

516.   SABLE, MARTIN H. (ed.), *A Guide to Latin American Studies*, Los Angeles: University of California, Latin American Center, 1967. 2 vols.

517.   ———, *Latin American Urbanization; A Guide to the Literature and Organization in the Field*, Los Angeles: University of California, Latin American Center, 1966. 117 pp.

518.   ———, *Master Directory for Latin America*, containing ten directories covering Organizations, Associations, and Institutions in the Fields of Agriculture, Business-industry-finance . . . Los Angeles: University of California, Latin American Center, 1965. 438 pp.

519.   SACHET, MARIE H., and FRANCIS R. FOSBERG, *Island Bibliographies: Micronesian Botany, Land Environment and Ecology of Coral Atolls, Vegetation of Tropical Pacific Islands*, Washington: National Academy of Sciences, National Research Council, 1955. 577 pp.

520. SAVORD, RUTH, and DONALD WATSON, *American Agencies Interested in International Affairs*, 5th ed., New York: Council on Foreign Relations, 1964. 200 pp.

521. *Scientific and Learned Societies of Great Britain*, London: Allen & Unwin, 1884 to date (publication suspended 1940–1950). Annual.

522. SCIENTIFIC COUNCIL FOR AFRICA, *Inventory of Economic Studies Concerning Africa South-of-the-Sahara; An Annotated Reading List of Books, Articles, and Official Publications*, Peter Adyl (ed.), London: Commission for Technical Cooperation in Africa South-of-the-Sahara, 1960. 301 pp. Publication No. 30, Joint Project No. 4.

523. SILBERMAN, BERNARD S., *Japan and Korea; A Critical Bibliography*, Tucson: University of Arizona Press, 1962. 136 pp.

524. SIMMS, RUTH P., *Urbanization in West Africa; A Review of Current Literature*, Evanston, Ill.: Northwestern University Press, 1965. 109 pp.

525. SNOW, PHILIP A. (comp.), *A Bibliography of Fiji, Tonga, and Rotuma*, Miami: 1969. 418 pp.

526. SOMMER, JOHN W., *Bibliography of African Geography*, Geographical Publications at Dartmouth No. 3, Hanover, N.H.: 1965. 139 pp.

527. SOUTH AFRICAN PUBLIC LIBRARY, CAPE TOWN, *A Bibliography of African Bibliographies, Covering Territories South-of-the-Sahara*, 4th ed., Cape Town: 1961. 79 pp.

528. STANFORD UNIVERSITY, THE HOOVER INSTITUTION, *United States and Canadian Publications on Africa*, Palo Alto. Annual since 1960.

529. TAYLOR, C. R. H., *A Pacific Bibliography*, 2nd ed., Oxford: Clarendon Press, 1965. 692 pp.

530. TREGONNING, K. G., *Southeast Asia: A Critical Bibliography*, Tucson, Ariz.: 1968.

531. TWENTIETH CENTURY FUND, *Survey of Tropical Africa; Select Annotated Bibliography of Tropical Africa*, Daryll Forde (ed.), New York: International African Institute, 1956.

532. UNION OF INTERNATIONAL ASSOCIATIONS, *International Institutions and International Organizations*, Brussels: 1963. 48 pp.

533. ———, *Yearbook of International Organizations*, 11th ed., Brussels: 1966. 1007 pp.

534. UNITED NATIONS, ECONOMIC COMMISSION FOR ASIA AND THE FAR EAST LIBRARY, *Asian Bibliography*, 1—. 1952. 2 numbers a year.

535. UNITED NATIONS EDUCATIONAL, SCIENTIFIC AND CULTURAL ORGANIZATION, *Directory of International Scientific Organizations*, Paris: 1953. 312 pp.

536. ———, *Directory of Reference Works Published in Asia*, P. K. Garde (comp.), Paris: 1956. 139 pp.

537. ———, FIELD SCIENCE COOPERATION OFFICE FOR LATIN AMERICA, *Scientific Institutions and Scientists in Latin America*, Montevideo: 1949 to date.

538. UNITED NATIONS, FOOD AND AGRICULTURE ORGANIZATION, *Bibliography on Land and Water Utilization and Conservation in Europe*, Rome: 1955. 347 pp.

539. U.S. ARMY, *Africa: Problems and Prospects: A Bibliographic Survey*, Washington: 1967. 226 pp.

540. U.S. ARMY NATICK LABORATORIES, *A Bibliography of Arid Lands Bibliographies*, Technical Report 68–27–ES, Natick, Mass.: 1967. 71 pp.

541. ———, *Desert Research: Selected References 1965–1968*, Natick, Mass.: 1969. 410 pp.

542. U.S. BUREAU OF THE CENSUS, *Bibliography of Social Science Periodicals and Monograph Series*, Foreign Demographic Analysis Division, Washington: Government Printing Office, Nos. 1–22, 1961–1965. (Foreign Social Science Bibliographies, Series P-92, Nos. 1–22.)

Detailed bibliographies of social science literature in periodicals and other serials in an individual country in the socialist group or using a language that is little known in the United States or considered difficult.

1. Rumania	1947–1960	1961	27 pp.
2. Bulgaria	1944–1960	1961	36 pp.
3. Mainland China	1949–1960		32 pp.
4. Republic of China	1949–1961	1961	24 pp.
5. Greece	1950–1961	1962	
6. Albania	1944–1961	1962	12 pp.

7. Hong Kong	1950–1961	1962	13 pp.
8. North Korea	1945–1961	1962	12 pp.
9. Republic of Korea	1945–1961	1962	48 pp.
10. Iceland	1950–1962	1962	10 pp.
11. Denmark	1945–1961	1963	111 pp.
12. Finland	1950–1962	1963	85 pp.
13. Hungary	1947–1962	1964	137 pp.
14. Turkey	1950–1962	1964	88 pp.
15. Norway	1945–1962	1964	59 pp.
16. Poland	1945–1962	1964	312 pp.
17. U.S.S.R.	1950–1963	1965	443 pp.
18. Yugoslavia	1945–1963	1965	152 pp.
19. Czechoslovakia	1948–1963	1965	129 pp.
20. Japan	1950–1963	1965	346 pp.
21. Soviet Zone of Germany	1948–1963	1965	190 pp.
22. Sweden	1950–1963	1965	83 pp.

543. U.S. DEPARTMENT OF COMMERCE, BUREAU OF FOREIGN COMMERCE, *A Guide to Foreign Business Directories*, Washington: 1955. 132 pp.

544. U.S. DEPARTMENT OF THE INTERIOR, DEPARTMENT LIBRARY, *Natural Resources in Foreign Countries; A Contribution toward a Bibliography of Bibliographies*, Bibliography No. 9, Mary Anglemyer (comp.), Washington: 1968. 113 pp.

545. U.S. LIBRARY OF CONGRESS, *Africa South of the Sahara: A Selected, Annotated List of Writings*, Helen F. Conover (comp.), Washington: 1963. 354 pp.

546. ———, *Agricultural Development Schemes in Sub-Saharan Africa: A Bibliography*, Ruth S. Freitag (comp.), Washington: 1963. 189 pp.

547. ———, *Antarctic Bibliography*, Washington: 1965.

548. ———, *Czechoslovakia, A Bibliographic Guide*, Washington: 1968. 157 pp.

549. ———, *A List of American Doctoral Dissertations on Africa*, Washington: 1962. 69 pp.

550. MILOJEVIC, BORIVOJE Z. (comp.), *Geography of Yugoslavia; A Selective Bibliography*, Washington: Reference Department, Slavic and East European Division, 1955. 79 pp.

551. U.S. LIBRARY OF CONGRESS, GENERAL REFERENCE AND BIBLIOGRAPHY DIVISION, *A Guide to Bibliographical Tools for Research in Foreign Affairs*, 2nd ed., Helen F. Conover (comp.), Washington: 1958. 145 pp.

552. ———, *North and Northeast Africa; A Selected, Annotated List of Writings, 1951–1957*, Helen F. Conover (comp.), Washington: 1957. 182 pp.

553. U.S. LIBRARY OF CONGRESS, HISPANIC FOUNDATION, *Handbook of Latin American Studies*, Cambridge: Harvard University Press, 1936–47; and Gainesville: University of Florida Press, 1948 to date. Annually.

554. ———, *National Directory of Latin Americanists*, Washington: 1966. 351 pp.

555. U.S. LIBRARY OF CONGRESS, ORIENTALIA DIVISION, *Southeast Asia; An Annotated Bibliography of Selected Reference Sources in Western Languages*, rev. ed., Cecil Hobbs (comp.), Washington: 1964. 180 pp.

556. U.S. LIBRARY OF CONGRESS, PROCESSING DEPARTMENT, *East European Accessions Index*, Vols. 1–10, 1951–1961, Washington: Government Printing Office.

557. U.S. NAVAL PHOTOGRAPHIC INTERPRETATION CENTER, *Antarctic Bibliography*, Washington: Government Printing Office, 1951. 147 pp.

558. WAGLEY, CHARLES (ed.), *Social Science Research on Latin America*, New York: Columbia University Press, 1964. 338 pp.

559. WEAVER, GERRY L., *Latin American Development: A Selected Bibliography*, Santa Barbara: ABC-Clio, 1969. 50 pp.

560. WEPSIEC, JAN, *A Checklist of Bibliographies and Serial Publications for Studies of Africa South-of-the-Sahara, Including Publications in the University of Chicago Libraries*, Chicago: University of Chicago, Committee on African Studies, 1966. 60 pp. Supplement, 1967.

561. WILBER, DONALD NEWTON, *Annotated Bibliography of Afghanistan*, 2nd ed., New Haven: Human Relations Area Files, 1962. 259 pp.

562. WISH, JOHN R., *Economic Development in Latin America: An Annotated Bibliography*. New York: Praeger, 1965. 144 pp.

563. YUAN, TUNG-LI, *Economic and Social Development of Modern China; A Bibliographical Guide,* New Haven: Human Relations Area Files, 1956. 130 pp.

564. ZIMMERMAN, IRENE, *A Guide to Current Latin American Periodicals; Humanities and Social Sciences,* Gainesville, Fla.: Kallman Publishing Co., 1961. 357 pp.

CHAPTER 7

Major Sources of Statistical Information

565. CLAWSON, MARION, and CHARLES L. STEWART, *Land Use Information: A Critical Survey of U.S. Statistics for Greater Uniformity*, Baltimore: Johns Hopkins Press, 1966.

566. CLAYTON, H. HELM (ed.), *World Weather Records*, Washington: Smithsonian Institution, 1927, 1937, 1947. Continued by *World Weather Records*, Washington: ESSA, 1965 to date.

567. *Editor and Publisher's Market Guide*, New York: 1924 to date. Annual. Statistics for cities and towns which have daily newspapers.

568. ENVIRONMENTAL SCIENCES SERVICE ADMINISTRATION, *Selected Guide to Published Climatic Data Sources Prepared by the U.S. Weather Bureau*, Silver Spring, Md.: 1969.

569. *Forest Industries—Buyers Guide and Yearbook Number*, Portland, Ore.: Miller Freeman Publications, 1962 to date. Annual.

570. *Information Please Almanac*, New York: Farrar, Straus & Giroux. Annual.

571. KENDALL, MAURICE G., and ALISON G. DOIG, *Bibliography of Statistical Literature*, New York: Hafner, 1962 to date.

572. *The Municipal Yearbook*, Chicago: The International City Managers' Association, 1934 to date. Annual.

169

573. PAN AMERICAN UNION, DEPARTAMENTO DE ESTADÍSTICA, *America en Cifras, 1965*, Washington: 1966. 3 vols.

574. PHILLIPPINES, BUREAU OF CENSUS and STATISTICS, *Facts and Figures about the Philippines*, Manila: 1965. 115 pp.

575. *Reader's Digest Almanac*, Pleasantville, N.Y.: The Reader's Digest Association. Annual.

576. STANFORD RESEARCH INSTITUTE, *Industrial Economics Handbook*, Menlo Park, Calif.: 1961. 5 vols. (U.S. economic statistics in graphs and tables; continual revision on loose-leaf sheets.)

577. STANFORD RESEARCH INSTITUTE, *Western Resources Handbook*, Menlo Park, Calif.: 1960. 4 vols. (Maps, graphs, tables; data for the 11 western states.)

578. *Statesman's Yearbook*, London and New York: Macmillan. Annual.

579. UNITED NATIONS, DEPARTMENT OF ECONOMIC and SOCIAL AFFAIRS, *Analytical Bibliography of International Migration Statistics*, Selected Countries, 1925–1950, New York: 1955. 195 pp.

580. UNITED NATIONS, FOOD AND AGRICULTURE ORGANIZATION, *Report on the 1960 World Census of Agriculture*, Rome: 1967.

581. ———,*Wood: World Trends and Prospects*, Rome: 1967. 130 pp.

582. ———, *World Crop Statistics: Area, Production and Yield*, 1948–64, Rome: 1966. 458 pp.

583. ———, *Yearbook of Fishery Statistics*, Rome. Annually since 1947.

584. ———, *Yearbook of Food and Agricultural Statistics; Part I, Production; Part II, Trade*, Rome. Annually 1947–1957. (Since 1958: *Production Yearbook*. Before 1947: *International Yearbook of Agricultural Statistics*.)

585. ———, *Yearbook of Forest Products Statistics*, Rome. Annually since 1947.

586. UNITED NATIONS, STATISTICAL OFFICE, *Demographic Yearbook*, New York. Annually since 1948.

587. ———, *Statistical Yearbook*, New York. Annually since 1948.

588. ———, *World Energy Supplies*, 1962–1965, New York: 1967.

589. ———, *Yearbook of International Trade Statistics*, New York: 1950 to date. Annual.

590. U.S. BUREAU OF THE BUDGET, *Federal Statistical Directory*, Washington: 1935 to date. Annual. (A guide to the statistical programs of the Federal Government.)

591. ———, *Statistical Services of the United States Government*, Washington. Issued irregularly.

592. U.S. BUREAU OF RECLAMATION, *Reclamation Project Data; A Book of Historical, Statistical and Technical Information on Reclamation Projects*, Washington: 1949. 489 pp.

593. U.S. DEPARTMENT OF AGRICULTURE, *Agricultural Statistics*, Washington. Annually since 1936. 1894–1935 in *Yearbook of Agriculture*. 1863–1893 in *Annual Report*.

594. ———, *Farm Population*, Washington. Annual.

595. ———, *Foreign Agricultural Trade of the United States*, Washington. Monthly with annual supplements.

596. U.S. DEPARTMENT OF COMMERCE, BUREAU OF THE CENSUS, *Census of Agriculture*, Washington. Irregular. 1925, 1950, 1954, 1959, 1964.

597. ———, *Census of Business*, Washington: 1929, 1933, 1939, 1948, 1958, 1963. Part of decennial census in 1929 and 1939.

598. ———, *Census of Housing*, Washington. Decennial 1951–1954 and 1960.

599. ———, *Census of Manufactures*, Washington. Included in decennial censuses 1810–1940; quinquennial censuses in 1905 and 1914; biennial 1919–1939; 1919, 1929, and 1939 as part of decennial census. Separate census in 1947, 1954, 1958, 1961.

600. ———, *Census of Mineral Industries*, Washington: 1939, 1954, 1958, 1963.

601. ———, *Congressional District Data Book*, Washington: 1963. 560 pp.

602. ———, *County and City Data Book*, Washington. Biennial.

603. ———, *Decennial Census of Population*, Washington: 1790 to date.

604. ———, *Directory of Federal Statistics for Local Areas*, Washington: 1967. Annual.

605. ———, *Foreign Commerce and Navigation of the United States*, Washington. Annually since 1865.

606. ———, *Guide to Industrial Statistics*, Washington: 1964. 60 pp.

607. ———, *Historical Statistics of the United States, Colonial Times to 1957*, Washington: 1960.

608. ———, *Pocket Data Book USA*, Washington: 1967 to date. Annual.

609. ———, *Statistical Abstract of the United States*, Washington: 1878 to date. Annual.

610. ———, *United States Exports of Domestic and Foreign Merchandise*, Washington. Monthly and annual.

611. ———, *United States Imports of Merchandise for Consumption*, Washington. Monthly and annual.

612. ———, *Waterborne Foreign Trade Statistics*, Washington. Monthly and annual.

613. U.S. DEPARTMENT OF COMMERCE, BUREAU OF FOREIGN AND DOMESTIC COMMERCE, *Foreign Commerce Yearbook*, Washington. Annually 1922–39; 1948 to date. Starts with 1933 (1934, 1940–1947 publications suspended).

614. U.S. DEPARTMENT OF THE INTERIOR, BUREAU OF MINES, *Minerals Yearbook*, Washington: annually since 1932. 1924–31 as *Minerals Resources of the United States*; 1880–1923 in *Annual Report of the Director, U.S. Geological Survey*.

615. ———, *Mineral Facts and Problems*, Washington: 1956 to date. Annually.

616. U.S. FEDERAL POWER COMMISSION, *Electric Power Statistics*, Washington. Monthly.

617. ———, *Statistics of Electric Utilities in the United States*, Washington. Annual.

618. ———, *Statistics of Natural Gas Companies*, Washington. Annual.

619. WASSERMAN, PAUL, et al, *Statistical Sources*, 2nd ed., Detroit: Gale Research, 1967. 387 pp.

620. WEAVER, JOHN C., and FRED K. LUKERMANN, *World Resource Statistics, A Geographic Sourcebook*, 2nd ed., Minneapolis: 1953.

621. WHITAKER, JOSEPH, *Almanack*, London: Whitaker, 1869.

622. *World Almanac and Book of Facts*, New York: World-Telegram. Annual.

CHAPTER 8
Map Sources

Maps containing the information you wish to show at a scale small enough to utilize in your research paper are difficult to find. However, you will save yourself a good deal of time if you are able to discover maps and graphs suited to your purpose.

Consult *A Basic Geographical Library* for the names of books and atlases pertinent to your project. For example, if you are working on a topic involving the historical geography of the United States, you may find useful maps in Paullin's *Atlas of the Historical Geography of the United States* or in the *Statistical Atlas* which accompanies many of the decennial censuses.

The lists found on this and succeeding pages contain sources from which you may obtain maps and other illustrative materials for your papers. If you have access to a map library, consult the reference librarian there.

A. General Lists, Indexes, Map Sources

American Congress on Surveying and Mapping, Box 470, Benjamin Franklin Station, Washington, D.C. 20044.

AMERICAN GEOGRAPHICAL SOCIETY, *Current Geographical Publications: Additions to the Research Catalogue of the American Geographical Society*, New York: 1938 to date.

ASSOCIATION OF AMERICAN GEOGRAPHERS, HIGH SCHOOL GEOGRAPHY PROJECT, *Sources of Information and Materials: Maps and Aerial Photographs,* Boulder, Colo.: 1970.

Bibliographie Cartographique Internationale, Paris: Armand Colin. Annual since 1948.

BLISS, RICHARD, "Classified Index to Maps in Petermann's Geographische Mittleilungen, 1855–1881," *Harvard University Bulletin,* III (April, 1883), 94–106.

BLISS, RICHARD, "Classified Index to the Maps contained in the Publications of the Royal Geographical Society, and in Associated Serials," *Harvard University Bulletin,* IV (January, 1886), 242–248.

CALIFORNIA STATE CHAMBER OF COMMERCE ECONOMIC DEVELOPMENT AND RESEARCH DEPARTMENT, *Sources of California Maps,* Sacramento: 1964. 26 pp.

CALIFORNIA STATE LIBRARY, *List of Printed Maps Contained in the Map Department,* Sacramento: A. J. Johnston, 1899. 43 pp.

CHAPIN, EDWARD L., *A Selected Bibliography of Southern California Maps,* with a foreword by Clifford H. MacFadden, Berkeley: University of California Press, 1953. 124 pp. Maps.

EARTH SCIENCE CURRICULUM PROJECT, *Selected Maps and Earth Science Publications for the States and Provinces of North America,* Englewood Cliffs, N.J.: Prentice-Hall, 1965. 48 pp.

"Foreign Topographic Mapping Agencies and Their Sales and Information Offices," *Surveying and Mapping,* 41(April–June, 1956), 212–216.

FREEMAN, OTIS E., "Maps and Their Use in Pacific Northwest History," *Pacific Northwest Quarterly,* 43(July, 1951), 203–210.

KOHL, J. G., *A Descriptive Catalogue of Those Maps, Charts and Surveys Relating to America Which are Mentioned in Volume III of Hakluyt's Great Work,* Washington, D.C.: Henry Polkinhorn, 1857. 86 pp.

KÜCHLER, A. W., and JACK McCORMICK, "Bibliography of Vegetation Maps of North America," *Excerpta Botanica,* Section B, 8(1967), 145–289.

LOCK, C B MURIEL, *Modern Maps and Atlases,* Hamden, Conn.: Archon Books, 1969. 619 pp.

NATIONAL GEOGRAPHIC SOCIETY, *Publication Order List,* Washington, D.C. Irregularly, unpaged.

ODYERS, CHARLOTTE E., "Federal Government Maps Related to Pacific

Northwest History," *Pacific Northwest Quarterly,* 38(July, 1947), 261–272.

"Report on Cartobibliographies," *The Canadian Cartographer,* 5(December, 1968), 149–153.

RISTOW, WALTER W., "A Half Century of Oil Company Road Maps," *Surveying and Mapping,* 24(December, 1964), 617–637.

ROYAL GEOGRAPHICAL SOCIETY, *New Geographical Literature and Maps,* London: 1951 to date.

SPECIAL LIBRARIES ASSOCIATION, 31 East 10th St., New York, N.Y. 10003. (Publishes a quarterly entitled *Geography and Map Division Bulletin.*)

SPECIAL LIBRARIES ASSOCIATION, GEOGRAPHY AND MAP DIVISION, *Map Collections in the United States and Canada,* New York: 1954. 170 pp.

TOBLER, W. R., *Maps of the United States: A Guide to What Maps Are Available, Where Obtainable and How to Order,* Seattle, Wash.: The Author, 1959. 32 pp.

U.S. DEPARTMENT OF THE INTERIOR, GEOLOGICAL SURVEY, *Maps of the United States,* Washington, D.C.: Irregular. Unpaged.

———, *Maps of the World,* Washington, D.C. Irregular. 1 p.

———, *State Maps,* Washington, D.C. Irregular. 22 pp.

U.S. GOVERNMENT PRINTING OFFICE, *Maps, Engineering, Surveying,* Price List 53, Washington, D.C. Annual.

U.S. LIBRARY OF CONGRESS, *A List of Geographical Atlases in the Library of Congress, with Bibliographical Notes,* by Philip Lee Phillips and Clara E. LeGear, Washington, D.C.: 1909–1963. 6 vols.

U.S. LIBRARY OF CONGRESS, GEOGRAPHY AND MAP DIVISION, "Annual Report of Acquisitions of the (Geography) and Map Division," *Quarterly Journal of the Library of Congress* (August, 1946 to 1961), (September, 1962 and 1963), (October, 1964), (July, 1965, 1966).

———, *Aviation Cartography, a Historico-Bibliographic Study of Aeronautical Charts,* 2nd ed., Walter W. Ristow (comp.). Washington, D.C.: 1960. 245 pp.

———, *Bibliography of Cartography.* (Consists of over 40,000 card references to literature about maps, map making, and map makers.) 24 reels of microfilm.

———, *Civil War Maps: An Annotated List of Maps and Atlases in the Map Collections of the Library of Congress*, Richard W. Stephenson (comp.), Washington, D.C.: 1961. 138 pp.

———, *A Descriptive List of Treasure Maps and Charts in the Library of Congress*, Richard S. Ladd (comp.), Washington, D.C.: 1964. 29 pp.

———, *A Guide to Historical Cartography*, Walter W. Ristow and Clara E. LeGear (comps.), Washington, D.C.: reprinted 1962. 22 pp.

———, *The Hotchkiss Map Collection; A List of Manuscript Maps, Many of the Civil War Period; Prepared by Major Jed Hotchkiss, and other Manuscripts and Annotated Maps in His Possession*, Clara E. LeGear (comp.), with a foreword by Willard Webb, Washington, D.C.: 1951. 67 pp.

———, *A List of Maps of America in the Library of Congress; Preceded by a List of Works Relating to Cartography*, Philip Lee Phillips, Washington, D.C.: 1901. 1137 pp.

———, *The Lowery Collection; A Descriptive List of Maps of the Spanish Possessions Within the Present Limits of the United States, 1502–1820*, edited with notes by Philip Lee Phillips, Washington, D.C.: 1912. 567 pp.

———, *Maps Showing Explorers' Routes, Trails and Early Roads in the United States; An Annotated List of Maps in the Map Collections of the Library of Congress*, Richard S. Ladd (comp.), Washington, D.C.: 1962. 137 pp.

———, *Marketing Maps of the United States; an Annotated Bibliography*, Walter W. Ristow (comp.), Washington, D.C.: 1958. 147 pp.

———, *Selected Maps and Charts of Antarctica; An Annotated List of Maps of the South Polar Regions*, Washington, D.C.: 1959. 193 pp.

———, *Three-Dimensional Maps*, 2nd ed., Walter W. Ristow (comp.), Washington, D.C.: 1964. 38 pp.

———, *United States Atlases; A List of National, State, County, City and Regional Atlases: in the Library of Congress*, Clara E. LeGear (comp.), Washington, D.C.: 1950, 1953. 2 vols.

U.S. NATIONAL ARCHIVES AND RECORDS SERVICE, *List of Cartographic Records of the General Land Office*, special list Number 19 compiled by Laura E. Kelsay, Washington, D.C.: 1964. 202 pp.

WHEAT, JAMES CLEMENTS, and CHRISTIAN F. BRUN, *Maps and Charts*

Published in America Before 1800: A Bibliography, New Haven: Yale University, 1969.

WINSOR, JUSTIN, "The Kohl Collection of Maps," *Harvard University Bulletin*, IV (January, 1886), 234–242.

YONGE, ENA L., "Regional Atlases: A Summary Survey," *The Geographical Review*, 52 (July, 1962), 407–432.

B. International and Foreign Sources

AUSTRALIA
Division of National Mapping, Derwent House, 22–34 University Avenue, Canberra City, A.C.T.

CANADA
Department of Mines and Technical Surveys, Geographical Branch, 601 Booth Street, Ottawa.
International Civil Aviation Organization, International Aviation Building, 1080 University Street, Montreal.

GERMANY
Reise und Verkehrslag, Postfach 8. 083. 7 Stuttgart.

NEW ZEALAND
Department of Scientific and Industrial Research, P. O. Box 8018, Wellington.

SCOTLAND
John Bartholomew and Son Ltd., 12 Duncan Street, Edinburgh 9.

SOVIET UNION
Glanoe Upravlenie Geodezii, i Kartografii, Moscow.
Telberg Book Co., 544 Sixth Avenue, New York, N.Y. 10011.

UNITED KINGDOM
Directorate of Military Survey, The War Office, London, SW1.
Directorate of Overseas Surveys, Department of Technical Cooperation, Kingston Road, Tolworth, Surrey.
Geographia Ltd., 114 Fleet Street, London, E.C.4.
Edward Stanford Ltd., 12–14 Long Acre, London, W.C.2.

WORLD
UNESCO Publications Center, 650 First Avenue, New York, N.Y. 10016.

C. United States Map Sources

1. Types of Maps Published by Government Agencies

TYPE	PUBLISHING AGENCY	AVAILABLE FROM
Aeronautical charts	Aeronautical Chart and Information Center, Coast and Geodetic Survey	Coast and Geodetic Survey
Boundary information:		
United States and Canada	International Boundary Commission	International Boundary Commission
United States and Mexico	International Boundary and Water Commission	International Boundary and Water Commission
Congressional districts	Bureau of the Census	Superintendent of Documents
Electrical facilities	Federal Power Commission	Superintendent of Documents
Geologic quadrangle maps	Geological Survey	Geological Survey
Geologic investigations maps	Geological Survey	Geological Survey
Geologic map of North America	Geological Survey	Geological Survey
Geophysical investigations maps	Geological Survey	Geological Survey
Ground conductivity	Federal Communications Commission	Superintendent of Documents
Highways:		
United States	Bureau of Public Roads	Superintendent of Documents
State and County	State Highway Departments	State Highway Departments

Type of map		
Historical:		
Reproductions from historical and military map collections	Library of Congress	Library of Congress
	National Archives	National Archives
Explorers' routes	Superintendent of Documents	Library of Congress
Treasure maps and charts (bibliography)	Superintendent of Documents	Library of Congress
Hydrographic information:		
Nautical charts of U.S. coastal waters	Coast and Geodetic Survey	Coast and Geodetic Survey
Charts of inland waters		
Great Lakes and connecting waters	Corps of Engineers, U.S. Lake Survey	Corps of Engineers, U.S. Lake Survey
River charts:		
Middle and Upper Mississippi River and Illinois Waterway to Lake Michigan	Corps of Engineers	Corps of Engineers
Lower Mississippi River	Corps of Engineers	Corps of Engineers
Missouri River	Corps of Engineers	Corps of Engineers
Ohio	Corps of Engineers, Cincinnati	Corps of Engineers
Foreign waters	U.S. Naval Oceanographic Office	U.S. Naval Oceanographic Office
Hydrologic investigations atlases	Geological Survey	Geological Survey
Indian reservations	Bureau of Indian Affairs	Bureau of Indian Affairs
Mineral resources maps and charts	Geological Survey	Geological Survey
Minor civil divisions	Superintendent of Documents	Bureau of the Census
National atlas of the United States	Superintendent of Documents	Various government agencies

1. Types of Maps Published by Government Agencies (*cont.*)

TYPE	PUBLISHING AGENCY	AVAILABLE FROM
National forests:		
Forest regions	Forest Service	Forest Service
National forest index	Forest Service	Superintendent of Documents
National parks:		
Topographic maps	Geological Survey	Geological Survey
National parks system	National Park Service	Superintendent of Documents
Natural gas pipelines	Federal Power Commission	Superintendent of Documents
Polar maps:		
Antarctic	Geological Survey	Geological Survey
	Coast and Geodetic Survey	Coast and Geodetic Survey
	U.S. Naval Oceanographic Office	U.S. Naval Oceanographic Office
Arctic	Coast and Geodetic Survey	Coast and Geodetic Survey
	U.S. Naval Oceanographic Office	U.S. Naval Oceanographic Office
Population distribution of the United States	Bureau of the Census	Superintendent of Documents
Railroad map of the United States	Corps of Engineers	Army Map Service (Texas)
Soil survey maps	Soil Conservation Service	Superintendent of Documents
Time zones:		
United States, Canada, and Mexico	Interstate Commerce Commission	Superintendent of Documents
World	U.S. Naval Oceanographic Office	U.S. Naval Oceanographic Office
Topographic map series of the United States:		

Indexes to published topographic maps for each state, Puerto Rico, and the Virgin Islands	Geological Survey	Geological Survey
Status indexes to aerial mosaics, aerial photography, and topographic mapping in the United States	Geological Survey	Geological Survey
Township plates (reproductions): Illinois, Indiana, Iowa, Kansas, Missouri, and Ohio	National Archives	National Archives
All other public land states	Bureau of Land Management	Bureau of Land Management
	Coast and Geodetic Survey	Coast and Geodetic Survey
United States base maps	Geological Survey	Geological Survey
	Other government agencies	Superintendent of Documents
Weather map	Weather Bureau	Superintendent of Documents
World maps	Army Map Service	Army Map Service (Washington, D.C.)
	Coast and Geodetic Survey	U.S. Naval Oceanographic Office
	U.S. Naval Oceanographic Office	Coast and Geodetic Survey

2. Government Publishing and Distributing Agencies

Army Map Service, Attn: 16230, Washington, D.C. 20315.

Army Map Service, San Antonio Field Office, Building 4011, Fort Sam Houston, Tex. 78234.

Bureau of the Census, Department of Commerce, Washington, D.C. 20233.

Bureau of Indian Affairs, Department of the Interior, Washington, D.C. 20242.

Bureau of Land Management, Department of the Interior, Washington, D.C. 20240.

Bureau of the Public Roads, Department of Commerce, Washington, D.C. 20235.

Coast and Geodetic Survey, Department of Commerce, Washington Science Center, Rockville, Md. 20852.

Corps of Engineers, 536 S. Clark Street, Chicago, Ill. 60605.

Corps of Engineers, P.O. Box 80, Vicksburg, Miss. 39180.

Corps of Engineers, 205 N. 17th Street, Omaha, Neb. 68102.

Corps of Engineers, P.O. Box 1159, Cincinnati, Ohio 45201.

Corps of Engineers, U.S. Lake Survey, 630 Federal Building, Detroit, Mich. 48226.

Federal Communications Commission, Washington, D.C. 20554.

Federal Power Commission, Washington, D.C. 20426.

Forest Service, Department of Agriculture, Washington, D.C. 20250.

Geological Survey, Department of the Interior, Washington, D.C. 20242.

Geological Survey, Federal Center, Denver, Colo. 80225. (West of the Mississippi.)

International Boundary Commission, U.S. and Canada, 441 G Street, N. W., Room 3810, Washington D.C. 20548.

International Boundary and Water Commission, United States and Mexico, United States Section, Mart Building, 4th Floor, El Paso, Tex. 79901.

Interstate Commerce Commission, Washington, D.C. 20423.

Library of Congress, Geography and Map Division, Washington, D.C. 20540.

National Archives, General Services Administration, Washington, D.C. 20408.

National Park Service, Department of the Interior, Washington, D.C. 20252.

Soil Conservation Service, Department of Agriculture, Washington, D.C. 20251.

State Department, Office of the Geographer, Washington, D.C. 20520.

State Highway Departments, state capitals.

Superintendent of Documents, Government Printing Office, Washington, D.C. 20402.

U.S. Air Force, Aeronautical Chart and Information Center, St. Louis, Mo. 63118.

U.S. Naval Oceanographic Office, U.S. Naval Supply Depot, 5801 Tabor Avenue, Philadelphia, Pa. 19111, or

U.S. Naval Oceanographic Distribution Office, Clearfield, Utah 84015.

Weather Bureau, Department of Commerce, Washington, D.C. 20235.

3. Commercial and Institutional Publishers of Maps, Globes, and Atlases

Aero Service Corporation, Division of Litton Industries, 4219 Van Kirk St., Philadelphia, Pa. 19135 (photogrammetry, molded plastic relief).

American Air Surveys, Inc., 907 Penn Ave., Pittsburgh, Pa. (topographic tax).

American Automobile Association, 1712 G Street, N.W., Washington, D.C.

American Geographical Society, Broadway at 156th Street, New York, N.Y. 10032.

American Map Company, Inc., 3 West 61st Street, New York, N.Y. 10023.

American Museum of Natural History, Central Park West and 79th Street, New York, N.Y. 10024.

Amman Photogrammetric Engineers, Inc., 931 Broadway, San Antonio, Tex. (oil cadastral).

G. W. Bromley & Company, 325 Spring St., New York, N.Y. 10013 (real estate, city).

Champion Map Corporation, P.O. Box 17435, Charlotte, N.C. 28211.

George F. Cram Company, Inc., P.O. Box 426, Indianapolis, Ind. 46206.

Denoyer-Geppert Company, 5235 Ravenswood Avenue, Chicago, Ill. 60640.

Diversified Map Corp., 9715 Page Boulevard, St. Louis, Mo. 63132.

Dolph Map Company, Inc., 430 N. Federal Highway, Ft. Lauderdale, Fla. 33301.

C. E. Erickson and Associates, 1715 Solano Avenue, Berkeley, Calif. (county maps of California).

Farquhar Transparent Globes, 5007 Warrington Ave., Philadelphia, Pa. (terrestrial, celestial).

General Drafting Company, Inc., Canfield Rd., Convent Station, N.J. (road).

Geographia Map Company, Inc., 220 West 42nd Street, New York, N.Y. 10036.

The Geographical Press, a Division of C. S. Hammond & Company (graphic depictions of surface features, geological). See C. S. Hammond & Co.

Geo-Physical Maps Inc., 521 Fifth Avenue, New York, N.Y. 10017.

H. M. Gousha Company, 2001 The Alameda, San Jose, Calif. (road).

Hagstrom Company, Inc., 311 Broadway, New York, N.Y. 10007.

C. S. Hammond & Co., Inc., 515 Valley Street, Maplewood, N.J. 07040.

Hearne Brothers, 25th Floor, National Bank Bldg., Detroit, Mich. 48226.

Howe Press of Perkins School for the Blind, Watertown, Mass. (Braille maps).

International Map Company Inc., 120 Liberty Street, New York, N.Y. 10013.

Jeppesen & Company, 8025 East 40th Avenue, Denver, Colo. 80207.

B. Orchard Lisle, Majestic Building, Fort Worth, Tex. (maps of oil fields).

Map Corp. of America, 316 Summer Street, Boston, Mass. 02210.

Metsker Maps, 111 South 10th St., Tacoma, Wash. 98402.

Midcontinent Map Company, 114 W. 3rd Street, Tulsa, Okla. (oil cadastral).

National Geographic Society, 1145 17th Street, N.W., Washington, D.C. 20036.

Thomas O. Nelson Company, Fergus Falls, Minn. 56536.

National Survey, Chester, Vt. 05143.

A. J. Nystrom & Company, 3333 Elston Avenue, Chicago, Ill. 60618.

Oxford University Press, Inc., 200 Madison Avenue, New York, N.Y. 10016.

Panoramic Studios, 179–189 W. Berks St., Philadelphia, Pa. (relief globes and models).

Price & Lee Company, 270 Orange Street, New Haven, Conn. 06511.

Erwin Raisz, 10 Emerson Place, Boston, Mass.

Rand McNally & Company, P.O. Box 7600, Chicago, Ill. 60680.

Renie Map Service, 1101 West Seventh Street, Los Angeles, Calif. (local area maps).

Replogle Globe Company, 1901 N. Narragansett Avenue, Chicago 29, Ill. 60629.

Sanborn Map Company, 629 5th Avenue, Pelham, N.Y. (insurance, atlases of urban areas).

Sidwell Studios, 28 West 240th North Avenue, West Chicago, Ill. 60185.

Thomas Bros. Maps, 550 Jackson Street, San Francisco, Calif. 94133.

The University of Chicago, Department of Geography, Chicago, Ill. 60615.

Weber Costello Company, Benefic Press, 1900 N. Narragansett Avenue, Chicago, Ill. 60639.

CHAPTER 9
Sources of Photographs

Photographs of all parts of the earth are available from a great variety of sources; some of them are listed on this and the following pages. Photographs taken at ground level may generally be obtained at no charge from chambers of commerce, development agencies, or from the public relations offices of transportation and travel agencies. Commercial photo agencies selling the right to use photos for a particular purpose have files containing hundreds of thousands of photographs of all parts of the earth. Most such companies are located in the New York and Los Angeles areas. Their fees are nominal: generally, five to ten dollars per photo.

Aerial photographs are more expensive and more difficult to obtain. The Map Information Office of the U.S. Geological Survey maintains records of aerial photographic coverage of the United States and outlying areas, based on reports of federal and state agencies and commercial companies. The Map Information Office will furnish you with data on sources of available photographs. State highway departments are additional sources of information on the availability of aerial photographs. Many of them publish index maps which show the parts of the state photographed in any given year. Offices of the Soil Conservation Service likewise maintain index maps of aerial photograph coverage for their local areas.

A. General Lists, Indexes, Photograph Sources

The Aerofilms Book of Aerial Photographs, London: Aerofilms Ltd., 4 Albemarle Street, 1965. 318 pp. (Photographs and lists of photographs for many parts of the world, especially British Isles.)

AMERICAN SOCIETY OF PHOTOGRAMMETRY, 1515 Massachusetts Avenue, N.W., Washington, D.C.

AMERICAN SOCIETY OF PHOTOGRAMMETRY, *The 1967 Services and Equipment Guide,* Falls Church, Va.: 1967. pp. 17–48, reprinted from *Photogrammetric Engineering* (January, 1967).

ASSOCIATION OF AMERICAN GEOGRAPHERS, HIGH SCHOOL GEOGRAPHY PROJECT, *Sources of Information and Materials: Maps and Aerial Photographs,* Washington, D.C.: 1970.

Bildquellen Handbuch, Der Wegweiser für Bildsuchende, Weisbaden: Verlag Chmielorz GMBH, 1961. (A directory of picture sources throughout the world.)

DURT, WAKEFIELD, JR., *Directory of Geoscience Films,* Washington, D.C.: American Geological Institute, 1962. 63 pp.

EDUCATIONAL MEDIA COUNCIL, *Educational Media Index,* New York: McGraw-Hill, 1964. (Includes films, slides, film strips, photographs.)

FRANKENBERG, CELESTINE G. (ed.), *Picture Sources,* New York: Special Libraries Association, 1964. 203 pp. (Gives both sources of pictures and reference books containing lists of potential picture sources.)

MILLER, BRUCE, *Sources of Free and Inexpensive Pictures for the Classroom,* Riverside, Calif.: The Author, Box 369, 1962. 30 pp. (Also charts and maps.)

NUNN, G. W. A. (ed.), *British Sources of Photographs and Pictures,* London: Cassell, 1952.

PARIS, BIBLIOTHÈQUE NATIONALE, *Répertoire des Collections Françaises de Documents Photographiques,* Paris: Editions de la Documentation Française, 1949. (Photographic sources of France; subject index.)

PROFESSIONAL PHOTOGRAPHERS OF AMERICA, INC., *Directory of Professional Photography*, 1961–62, Milwaukee, Wis.: 1962. (Lists photographers by geographical areas.)

Remote Sensing of the Urban Environment: A Selected Bibliography, Monticello, Ill.: Council of Planning Librarians, 1969. (Exchange Bibliography, 69.)

SPECIAL LIBRARIES ASSOCIATION, *Picturescope: Newsletter of Picture Division of the Special Libraries Association*, New York: 1953 to date. Quarterly. (Presents frequent descriptions of picture collections.)

STEINER, DIETER, *List of Bibliographers on Photo Interpretation*, Zurich: The Author, 1968. (Arbeiten Ausdem Geography Institute der Universitat Zurich, Ser. A, No. 208.)

STOKES, ISAAC N. P., and D. C. HASKELL, *American Historical Prints: Early Views of American Cities*, New York: New York Public Library, 1933.

STONE, KIRK H., "World Air Photo Coverage, 1960," *Photogrammetric Engineering*, 27 (1960), 214–227.

UNITED NATIONS, *Proceedings of the United Nations Seminar on Aerial Survey Methods and Equipment*, Bangkok: 1960. (Information on Asiatic sources of photographs.)

U.S. DEPARTMENT OF THE INTERIOR, GEOLOGICAL SURVEY, *Aerial Photographic Reproductions*, Washington, D.C. Irregular. Unpaged. (Describes USGS Photography.)

U.S. DEPARTMENT OF THE INTERIOR, GEOLOGICAL SURVEY, *Status of Aerial Photography*, Washington, D.C. Annual. (A map of the United States showing photographed areas reported by government agencies and commercial firms.)

U.S. LIBRARY OF CONGRESS, *Guide to the Special Collections of Prints and Photographs in the Library of Congress*, Paul Vanderbilt (comp.), Washington, D.C.: 1955.

U.S. LIBRARY OF CONGRESS, *Image of America: Early Photography, 1839–1900*, Washington, D.C.: 1957. (Catalog of exhibit held in Library of Congress, 1957.)

U.S. LIBRARY OF CONGRESS, *Selective Checklist of Prints and Photographs Recently Catalogued and Made Available for Reference*, Lots 2280–2984, 1949; Lots 3443–4120, 1950; Lots 4121–4801, 1954.

U.S. LIBRARY OF CONGRESS, PHOTODUPLICATION SERVICE, *Pictorial Americana*, 2nd ed., Milton Kaplan (comp.), Washington, D.C.: 1955. (A select list of photographic negatives; prints may be ordered from this catalog.)

B. United States Sources

1. Governmental

Aeronautical Chart and Information Center, Photographic Research and Services Division, 2nd and Arsenal Streets, St. Louis, Mo.

Agricultural Stabilization and Conservation Service, U.S. Department of Agriculture, Eastern Laboratory, 45 South French Broad Avenue, Asheville, N.C. 28801.

Agricultural Stabilization and Conservation Service, U.S. Department of Agriculture, 2505 Parleys Way, Salt Lake City, Utah 84109 (west of the Mississippi.)

American Society of Photogrammetry, 105 North Virginia Avenue, Falls Church, Va. 22046.

Bureau of Land Management, Department of the Interior, Washington, D.C. 20242.

Forest Service, U.S. Department of Agriculture, Washington, D.C. 20250.

Soil Conservation Service, U.S. Department of Agriculture, Federal Center Building, East-West Highway and Belcrest Road, Hyattsville, Md. 20781.

U.S. Army Photographic Agency, Room 5A–470, The Pentagon, Washington, D.C.

U.S. Geological Survey, Map Information Office, Washington, D.C. 20242.

U.S. Library of Congress, Prints and Photographs Division, Washington, D.C. 20540.

U.S. National Archives, Audio-Visual Records Branch, 9th and Constitution Avenues, N. W., Washington, D.C.

U.S. Naval Photographic Center, United States Naval Air Station, Anacostia, D.C.

2. Private

Abrams Aerial Survey Corp., 124 N. Larch, Lansing, Mich. 48905.

Adastra Survey Group, 41–45 Vickers Ave., Mascot, Sydney, N.S.W., Australia.

Aerial Mapping Co., Box 1957, Boise, Idaho 83701 or Box 7192, Phoenix, Ariz. 85011.

Aerial Mapping Co. of Oregon, 300 Education Center Bldg., 220 S. W. Alder St., Portland, Ore. 97204.

Aerial Map Service Co., 1016 Madison Ave., Pittsburgh, Pa. 15212.

Aerial Photography, McCullaugh Studio, 73 Simcoe St., Toronto, Ontario, Canada.

Aerial Photography Co., 741 E. McDowell Rd., Phoenix, Ariz. 85006.

Aerial Photo Service, Inc., 1127 North Cincinnati, Tulsa, Okla. 74106.

Aerial Photo & Survey Co., 780 S. Ridge Road, Lake Forest, Ill. 60045.

Aerial Surveys, Inc., 4614 Prospect Ave., Cleveland, Ohio 44103.

Aero Exploration Company, P. O. Box 7068, Tulsa, Okla.

Aero Exploration Vo/s., Flughafen, 6 Frankfurt am Main, (W.) Germany.

Aero Service Corp., Division of Litton Industries, 4219 Vankirk St., Philadelphia, Pa. 19135.

Aircraft Operating Co., (Aerial Surveys) Ltd., 23 Roger St., Selby, Johannesburg, Rep. of South Africa.

Air-Photo Co., Inc., 1901 Embarcadero Road, Palo Alto Airport, Palo Alto, Calif. 94303.

Air Photographics, Inc., 2417 Linden Lane, Silver Spring, Md. 20910.

Air Photo Surveys, Inc., P. O. Box 491, Grand Junction, Colo. 81502.

Air Survey Corp., 11445 Newton Square South, Reston, Va. 22070.

Airphoto Services Co., P.O. Box 834, Caldwell, N.J. 07006.

American Air Surveys, Inc., P.O. Box 491, Grand Junction, Colo. 81502.

Ammann-International, Div. of Knox, Bergman, Shearer & Associates, Inc., 223 Tenth St., San Antonio, Tex. 78215.

Atlantic Aerial Surveys, Inc., 803 Franklin St. S.E., Huntsville, Ala. 35801.

Michael Baker, Jr., Inc., P.O. Box 111, Rochester, Pa. 15074. 3957–67 Northview Drive, North Side Sta., P.O. Box 9997, Jackson, Miss. 39206.

Carl M. Berry, Consulting Engineering, South Annex, Boeing Field International, Seattle, Wash. 98108.

Bucher & Willis, Consulting Engineers, Planners & Architects, P. O. Box 1287, Salina, Kan. 67401.

Campbell Photo & Printing Service, 1328 I St., N.W., Washington, D.C. 20005.

Carto-Photo Corp., 520 Conger St., Eugene, Ore. 97402.

Cartwright Aerial Surveys, Inc., Municipal Airport, Sacramento, Calif. 95822.

Chicago Aerial Survey, 10265 Franklin Avenue, Franklin Park, Ill. 60131.

H. G. Chickering, Jr., Consulting Photogrammetrist, Inc., 1190 Seventh Ave. West, Eugene, Ore. 97402.

Christianson Enterprises, 5123 Carney Rd., Calgary, Alberta, Canada.

Clark Aerial Survey Corp., 3444 Highland Road (M–59), Pontiac, Mich. 48054.

Continental Engineers, Inc., 9625 West Colfax Avenue, Denver, Colo. 80215.

Cooks Aerial Photographic Surveys, Route 3, Alcoa, Tenn. 37701.

L. M. Dearing Assoc., Inc., 12345 Ventura Blvd., Suite R, Studio City, Calif. 91604.

Delano Photographics & Western Mapping Co., 1536 S. E. 11th Avenue, Portland, Ore. 97214.

Dillon Aerial Photography, 10 N. E. 3rd St., Fort Lauderdale, Fla. 33301.

Edgerton, Germeshausen & Grier, Inc., 160 Brookline Ave., Boston, Mass. 02215.

Fairchild Aerial Surveys, Inc., 10 Rockefeller Plaza, New York, N.Y. 10020.

Gates Aerial Surveys, 14 N. Main St., Spring Valley, N.Y. 10977.

Geophoto Services, Inc., P.O. Box 22293, Denver, Colo. 80222.

Greenwood & Assoc., 6410 Freeport Blvd., Sacramento, Calif. 95822.

Hammon, Jensen & Wallen, 8407 Edgewater Dr., Oakland, Calif. 94621.

Harris Aerial Surveys, Spring Branch Rd., Mountain Home, Ark. 72653.

Henderson Aerial Surveys, Ltd., 5125 W. Broad St., Columbus, Ohio 43228.

Clair A. Hill & Assoc., Consulting Engineers, P.O. Box 2088, Redding, Calif. 96002.

Mark Hurd Aerial Surveys, Inc., 345 Pennsylvania Ave. S., Minneapolis, Minn. 55426.

International Aerial Mapping Co., 127 International Dr., San Antonio, Tex. 78216.

Itek Corporation, Lexington, Mass. 02173.

Albert C. Jones Assoc., 223 High St., Mount Holly, N.J. 08060.

L. Robert Kimball, Consulting Engineers, 615 W. Highland Ave., Ebensburg, Pa. 15931.

Kucera & Assoc., Inc., 7000 Reynolds Rd., Mentor, Ohio 44060; 3835 Elm St., Denver, Colo. 80207; Lakeland Municipal Airport, Lakeland, Fla. 38803.

Latady Development Co., Inc., 38 W. Nippon St., Philadelphia, Pa. 19119.

Lewis-Dickerson Assoc., P.O. Box 576, Watertown, N.Y. 13601.

Limbaugh Engineers, Inc., 3125 Carlisle Blvd. N.E., P.O. Box 335, Albuquerque, N. Mex. 87103.

Lockwood, Kessler & Bartlett, Inc., One Aerial Way, Syosset, N.Y. 11791.

Lockwood Mapping, Inc., 580 Jefferson Rd., Rochester, N.Y. 14623.

Thomas M. Lowe, Jr. & Assoc., Inc., 1920 Monroe Drive, Atlanta, Ga. 30324.

The Lufkin Rule Co., 1730 Hess St., Saginaw, Mich. 48601.

K. B. MacKichan & Assoc., 211 N. Fifth St., Grand Forks, N.Dak. 58201.

Mapco, Inc., P.O. Box 1926, Ormond Beach, Fla. 32074.

Maps, Inc., Dundalk Marine Terminal, Baltimore, Md. 21222.

Metrex Aerial Surveys Co., 5020 Eagle Rock Blvd., Los Angeles, Calif. 90041.

Mississippi Aerial Photography, Inc., P.O. Box 4607, Fondren St., Jackson, Miss. 39216.

Moore Survey & Mapping Corp., 29 Grafton Circle, Shrewsbury, Mass. 01545.

National Aerial Surveys Inc., 421 Hawthorne Land, Charlotte, N.C. 28204.

Olympus Aerial Surveys Inc., 429 S. 5th East, Salt Lake City, Utah 48102.

Owen and White, Inc., P.O. Box 66396, Baton Rouge, La. 70806.

Pacific Air Industries Inc., 725 E. Third St., Long Beach, Calif. 90812.

Park Aerial Surveys, Inc., 606 Harding Ave., Louisville, Ky. 40221.

Pathfinder Air Surveys Ltd., 3323 Carling Ave., Ottawa 14, Ontario, Canada.

Peninsula Assoc., 1447 E. Bayshore, Palo Alto, Calif. 94303.

Photographic Surveys Inc., 8375 Bougainville St., Montreal 9, Quebec, Canada.

Photo-Tech, Inc., 330 Flume St., Chico, Calif. 95926; 523 E St., Marysville, Calif.

Photo-Topo Engineers, 7423 Crenshaw Blvd., Los Angeles, Calif. 90043.

Pictorial Crafts Inc., 1037 W. Base Line, San Bernardino, Calif. 92410.

Piedmont Aerial Surveys, Inc., P.O. Box 9177, Greensboro, N.C. 27408.

E. S. Preston & Assoc., Ltd., 939 Goodale Blvd., Columbus, Ohio 43212.

Quinn & Assoc., Inc., 835 Glenside Ave., Wyncote, Pa. 19095.

Rader & Assoc., The First National Bank Bldg., Miami, Fla. 33131.

Harman Rasnow & Assoc., 21318 Dumetz Rd., Woodland Hills, Calif. 91364.

Recon Assoc., 3020 Cedarwood Lane, Falls Church, Va. 22042.

Robinson Aerial Surveys, Inc., 43 Sparta Ave., Newton, N.J. 07860.

Sanborn Map Co., Inc., 629 Fifth Avenue, Pelham, N.Y. 10803.

San Diego Aerial Surveys, Inc., 8092 Engineering Rd., San Diego, Calif. 92111.

James W. Sewall Co., 147 Center St., Old Town, Me. 04468.

Shaw Photogrammetric Services Ltd., 285 Richmond Rd., Ottawa 3, Ontario, Canada.

Sholtz & Assoc., Inc., 5430 E. Beverly Blvd., Los Angeles, Calif., 90022.

Sidwell Studio, Inc., 28 W. 240 North Ave., West Chicago, Ill. 60185.

Spence Air Photos, 3614 West Pico Blvd., Los Angeles, Calif.

Sunderland Aerial Photographs Pacific Resources, Inc., P.O. Box 2416 Airport Station, Oakland, Calif. 94614.

Surdex Corp., St. Louis Air Park, Chesterfield, Mo. 63017.

Technology Inc., 7400 Colonel Glenn Hwy., Dayton, Ohio 45431.

Teledyne, Inc., Geotronics Div., 1000 S. Magnolia Ave., Monrovia, Calif. 91016.

Texas Instruments Inc., Science Services Div., P.O. Box 5621, Dallas, Tex. 75222.

R. M. Towill Corp., 233 Merchant St., Honolulu, Hawaii 96813.

R. M. Towill, Inc., 612 Howard St., San Francisco, Calif. 94105.

Underwood & Parker, Inc., 1710 W. Ninth St., Greeley, Colo. 80631.

Walker & Whiteford, Inc., 310 Prefontaine Bldg., Seattle, Wash. 98104.

Wallace Aerial Surveys, 1415 College Ave., P.O. Box 476, South Houston, Tex. 77587.

Western Aerial Contractors Inc., Mahlon Sweet Airport, Eugene, Ore. 97401.

Western Air Maps, Inc., 13001 W. 95th St., P.O. Box 5186, Lenexa, Kans. 66215.

The Ken R. White Co., Falcon Air Maps Div., 1567 Marion St., Denver, Colo. 80218.

Clyde E. Williams & Assoc., Inc., 1902 Sheridan Ave., South Bend, Ind. 46628.

C. Foreign Sources

BELGIUM
Institute Geographique Militaire, Brussels, Belgium.

BRAZIL
Servicos Aerofotogrammetricos, Cruzeiro Do Sul, S. A., Ave. Almirante
Frontin 381, Rio de Janeiro, Brazil ZC–22.

CANADA
Aerial Photography, McCullaugh Studio, 73 Simcoe Street, Toronto,
Ontario.
Aero Photo Inc., 1975 Charest Blvd., West, St. Fog, Quebec 10, P. Q.
Department of Mines and Technical Surveys, National Air Photo
Library, 615 Booth Street, Ottawa, Ontario.
Orhan's Reproductions and Photomapping Ltd., 718–8th Ave. S. W.,
Calgary, Alberta.
Spartan Air Services Ltd., 2117 Carling Ave., Ottawa 13, Ontario.

DENMARK
Royal Danish Geodetic Institute, Proviantgaarden, Rigsdagsgaarden 7,
Copenhagen K.

FINLAND
Finnmap Engineering Co., Kiviaidankatu 9, Helsinki.
National Board of Survey, Kirkokk 3, Helsinki.

FRANCE
Institut Géographique National Centre de Documentation Photo-
graphique, 2 Avenue Pasteur, Saint-Mande (Seine).
Institut Géographique National Serv. des cessions des cartes l'Ingenieur
en chef Géographe, 107 Rue la Boetie, Paris.

GERMANY
Aero-Exploration Vo/s, Flughafen, 6 Frankfurt a/m.
Hansa Luftbild GMBH, Postfach 1153, 44 Munster.
Luftbildtechnik GMBH, 1 Berlin (West), 42 Tempelhof Airport.
Photogrammetrie GMBH, Mauerkircherstrasse 1, Munich 27.

INDIA
Air Survey Co. Private Ltd., Dum Dum, Calcutta.

ITALY
EIRA, 44 Via Carlo Bini, Bellosguardo 4, Florence.
Instituto Geografico Militare, Florence.

ISRAEL
Ministry of Labour, Survey of Israel, P.O. Box 2730, Tel-Aviv.
Orient Press Photo Company, Dizengof Str., 97, Tel-Aviv. (Middle East
air photos.)

JAPAN
Asia Air Survey Co., Ltd., No. 594, 3-Chome, Tsurumaki-Cho Sefagaya-Ku, Tokyo.
Geographical Survey Institute, Tokyo.

LATIN AMERICA
Pan American Union, Natural Resources Unit, Washington, D.C. (Publish an index map of aerial photography.)

MALAYA
Survey Department, Gurney Road, Kuala Lumpur.

MEXICO
IFEX-Geotecnica, S. A., Lafragua #4–101, Mexico, D. F.

NETHERLANDS
International Institute for Aerial Survey and Earth Sciences, 3 Kanaalweg, Delft.
KLM Aerocarto, n. v., Schipol Airport, Delft.

NORWAY
Geographical Survey of Norway, St. Olavsgt. 32, Oslo.
Wideroe's Flyveselskap A/s, Kr. Augustsgate 19, Oslo.

PORTUGAL
Instituto Geografico e Cadastral, Praça de Estrela, Lisbon 2.

SOUTH AFRICA
Aircraft Operating Co., Ltd., 23 Roger Street, Selby, Johannesburg.
Trigonometric Survey Office, Pretoria.

SOVIET UNION
SOVFOTO, 24 West 45th Street, New York, N.Y.

SWEDEN
Geographical Survey Office of Sweden, Hantverkargatan 29, Stockholm K.

SWITZERLAND
Eidgenössische Landestopographie, Seftigenstrasse 264, 3084 Wabern, Bern.
Swissair Photo AG, 27 Sumatra St., 8006 Zurich.

UNITED KINGDOM
Aerofilms Limited, 4 Albemarle Street, London, W. 1., (Aerial photos principally British Isles, Africa, Middle East, Ceylon, Fiji, Cyprus, Burma, and Thailand.)
Aero Stills Ltd., 5 Kensington Church St., London W.8.

196

Air Ministry, Whitehall, London S.W.1.
BKS Air Survey Ltd., Cleeve Road, Leatherhead, Surrey.
Fairey Surveys Ltd., Reform Road, Maidenhead, Berkshire.
Hunting Surveys Ltd., 6 Elstree Way, Boreham Wood, Herts.
Meridian Airways Ltd., Commerce Way, Lancing, Sussex.
Overseas Geological Survey, Kingston Road, Tolworth, Chessington, Surrey.

WORLD
United Nations, Photographic Department, Room 994, New York, N.Y.

CHAPTER 10

A Selected List of Periodicals Used by Geographers

(Key to abbreviations used in list of periodicals.)

a. = annually
bi-m. = bimonthly
fortn. = fortnightly
irreg. = irregularly
m. = monthly
q. = quarterly
s-a. = semiannually
w. = weekly
3/yr. = three times a year
6/yr. = six times a year

A. Geography

Akademiia Nauk SSSR. 1951. bi-m. *Izvestiia Seriia Geografiches Kaia.*

Annales de Géographie. 1891. 6/yr. Librairie Armand Colin, 103 bd. St. Michel, Paris (5ᵉ), France.

Area. 1969. q. Institute of British Geographers, 1 Kensington Gore, London S.W. 7, England.

Association de Géographes Français, *Bulletin.* 1924. bi-m. Association de Géographes Français, 191 rue St. Jacques, Paris (5°), France.

ASSOCIATION OF AMERICAN GEOGRAPHERS, *Annals.* 1911. q. Association of American Geographers, Central Office, 1146 16th St., N.W., Washington, D.C. 20036.

ASSOCIATION OF PACIFIC COAST GEOGRAPHERS, *Yearbook.* 1935. a. John Gaines, Department of Geography, San Fernando Valley State College, Northridge, Calif. 91324.

Australian Geographer. 1929. s-a. Sydney University Press, University of Sydney, New South Wales, Australia.

Australian Geographical Studies. 1963. s-a. Institute of Australian Geographers, University of Melbourne, Parkville N.2, Australia.

Cahiers de Géographie de Quebec. 1956. 3/yr. University Lavel, Institut de Géographie, Quebec, Canada.

California Geographer. 1960. a. Robert W. Durrenberger, San Fernando Valley State College, Northridge, Calif. 91324.

Canadian Geographer/Géographe Canadien. 1951. q. University of Toronto Press, University of Toronto 5, Canada.

Canadian Geographical Journal. 1930. m. Royal Canadian Geographical Society, 488 Wilbrod St., Ottawa 2, Canada.

Deccan Geographer. 1962. s-a. Deccan Geographical Society, Mohd. Inamullah, 120/a Nehru Nagar East, Secunderabad, A.P., India.

East Lakes Geographer. (Association of American Geographers-East Lakes Division.) 1965. a. Department of Geography, Ohio State University, Columbus, Ohio 43210.

Economic Geography. 1925. q. Clark University, Worcester, Mass. 01610.

Erde. 1839. 4/yr. Walter De Gruyter & KCo., Genthiner Str. 13, Berlin 30, Germany.

Erdkunde. 1947. q. Institut, Franziskarerstr. 2, 53 Bonn, Germany.

Focus. 1950. m. (Sept.–June). American Geographical Society, Broadway at 156th St., New York, N.Y. 10032.

Geografia. 1962. s-a. Pakistan Institute of Geography, 1048 P.I.B. Colony, Karachi 5, Pakistan.

Geografiska Annaler, Series A: Physical Geography. 1919. q. Department of Geography, Uppsala 8, Sweden.

Geografiska Annaler, Series B: Human Geography. 1919. s-a. Drotting-gatan 120, Stockholm Va., Sweden.

Geographica Helvetica. 1946. q. Kuemmerley & Frey, Geographischer Verlag, Bern, Switzerland.

Geographical Analysis. 1969. q. Department of Geography, Ohio State University, Columbus, Ohio 43210.

Geographical Journal. 1893. q. Royal Geographical Society, 1 Kensing-ton Gore, London S.W.7, England.

Geographical Magazine. 1935. m. Oldhams Press Ltd., 96 Long Acre, London W.C.2, England.

Geographical Review. 1916. q. American Geographical Society, Broad-way at 156th St., New York, N.Y. 10032.

Geographical Review of India. 1933. q. Geographical Society of India, Calcutta 19, India.

Geographical Review of Japan/Chirigaku Hyoroa. 1925. m. Japan Pub-lications Trading Co., Ltd., C.P.O. Box 722, Tokyo, Japan.

Geographische Berichte. 1956. 4/yr. VEB Hermann Haack Geogra-phisch-Kartographie Anstalt, Justus-Perthes-Str. 3/9, Gotha, Ger-many.

Geographische Rundschau. 1949. m. Georg Westermann Allee 66, 33 Braunschweig, W. Germany.

Geographische Zeitschrift. 1905. q. Franz Steiner Verlag GmbH., Bahn-hofstr. 39, Wiesbaden, W. Germany.

Geography. 1901. q. George Philip & Son Ltd., 98 Victoria Rd., Willes-den, London, N.W.10, England.

IGU Newsletter/UGI Bulletin de Nouvelles. 1950. s-a. International Geographical Union, Secretariat, 10 Blumlisalpstr., 8006 Zurich, Switzerland.

Indian Geographical Journal. 1926. q. Indian Geographical Society, Uni-versity Centenary Bldgs., Chepauk, Madras 5, India.

INSTITUTE OF BRITISH GEOGRAPHERS, *Transactions.* 1935. s-a. George Philip & Son Ltd., 98 Victoria Rd., Willesden, London N.W.10, England.

Jobs in Geography. 1958. m. Association of American Geographers, 1146 16th St., N.W., Washington, D.C. 20036.

Journal of Geography. 1902. m. (Sept.–May.) National Council for Geographic Education, 7400 Augusta St., River Forest, Ill. 60305.

Journal of Tropical Geography. 1953. s-a. University of Singapore & University of Malaya, Department of Geography, Kuala Lumpur, Malaysia.

Leningradski Universitet, Vestnik: Seruja Geologii/Geografii. 1946. q. Mendeleyeoskaya lin 3/5, Leningrad V-164, USSR.

National Geographical Journal of India. 1955. q. Banaras Hindu University, Department of Geography, Varanasi 5, India.

National Geographic Magazine. 1888. m. 17th and M Sts., N.W., Washington, D.C. 20036.

New Zealand Geographer. 1945. s-a. University of Canterbury, Christchurch, New Zealand.

New Zealand Geographical Society, *Record.* 1946. s-a. New Geographical Society. University of Canterbury, Christchurch, New Zealand.

Oriental Geographer. 2/yr. East Pakistan Geographical Society, Department of Geography, University of Dacca, Dacca, Pakistan.

Pacific Viewpoint. 1960. s-a. Victoria University of Wellington, Geography Department, Box 196, Wellington, New Zealand.

Pakistan Geographical Review, 1942. s-a. University of the Punjab, Department of Geography, New Camjsus, Lahore, W. Pakistan.

Petermanns Geographische Mitteilungen. 1855. q. Justus-Perthe-Str. 3, Gotha, Germany.

Philippine Geographical Journal. 1953. q. Philippine Geographical Society. National Science Development Board, Manila, Philippine Islands.

Professional Geographer. 1949. bi-m. 1146 16th St., N.W., Washington, D.C. 20036.

Progress in Geography. 1969. a. St. Martin's Press, 175 Fifth Ave., New York, N.Y. 10010.

Revista Brasileira de Geografia. 1939. 4/yr. Conselbo Nacional de Geografia, Avenida Beira Mar 436, Rio de Janeiro, Brazil.

Revue de Géographie de Montreal. 1947. s-a. (Formerly *Revue Canadienne de Géographie.*) Université de Montréal, Département de Géographie, C.P. 6128, Montreal 3, P.Q., Canada.

Revue de Géographie et de Géologie Dynamique. 1928. q. Masson & Cie, 120 bd. Saint-Germain, Paris (6e), France.

Rivista Geografica Italiana. 1894. q. Piazza Independenza 29, Florence, Italy.

Scottish Geographical Magazine. 1885. 3/yr. 10 Randolph Crescent, Edinburgh 3, Scotland.

SOCIETÀ GEOGRAFICA ITALIANA, *Bollettino.* 1863. m. Società Geografica Italiana, Via della Navicella 12, Rome, Italy.

Southeastern Geographer. (Association of American Geographers, Southeastern Division.) 1961. a. Department of Geography, University of North Carolina, Chapel Hill, N.C.

Soviet Geography: Review and Translation, 1960. m. (Sept.–June.) American Geographical Society, Broadway at 156th St., New York, N.Y. 10032.

Tijdschrift Voor Economischa en Sociale Geografie. 1910. bi-m. Royal Dutch Geographical Society, Drukkerij van Waesberge, Hooge-werff en Richards N.V., Banierstraat 1, Rotterdam 1, Netherlands.

Walkabout. 1934. m. Australian National Travel Association, 18 Collins St., Melbourne, Australia.

Zeitschrift Fuer den Erdkundeunterricht. 1949. m. Volk und Wissen Volkseigener Verlag, Lindenstr. 54a, 108 Berlin, E. Germany.

B. Natural Sciences

Advancement of Science. 1939. m. J. M. Robertson. British Association for the Advancement of Science, 3 Sanctuary Bldgs., 20 Great Smith St., London S.W.1, England.

Agricultural Engineering. 1920. m. 420 Main St., St. Joseph, Mich., 49085.

Agricultural Meteorology. 1964. bi-m. Elsevier Publishing Co., Box 211, 335 Jan van Galenstraat, Amsterdam, Netherlands.

Agronomy Journal. 1907. bi-m. American Society of Agronomy, 2702 Monroe St., Madison, Wis. 53711.

AIR POLLUTION CONTROL ASSOCIATION, *Journal.* 1951. m. 4400 Fifth Ave., Pittsburgh, Pa. 15213.

American Forests. 1895. m. American Forestry Association, 919 Seventeenth St., N.W., Washington, D.C. 20006.

AMERICAN GEOPHYSICAL UNION, *Transactions.* 1919. q. 1515 Massachusetts Ave., N.W., Washington, D.C. 20036.

American Journal of Science. 1818. 10/yr. Yale University, New Haven, Conn. 06520.

AMERICAN METEOROLOGICAL SOCIETY, *Bulletin.* 1920. m. (Sept.–June.) 45 Beacon St., Boston, Mass. 02108.

American Scientist. 1911. q. 33 Witherspoon Street, Princeton, N.J. 08540.

Archiv für Meteorologie, Geophysik und Bioklimatologie. Series A: Meteorologie und Geophysik; Series B: Allgemeine und biologische Klimatologie. 1948. s-a. Springer-Verlag, Molkerbastel 5, Vienna 1, Austria.

Arctic. 1941. q. 3458 Redpath St., Montreal 25, Canada.

Arctic and Alpine Research. 1969. q. Institute of Arctic and Alpine Research, University of Colorado, Boulder, Colo. 80304.

Bulletin of the Atomic Scientists. 1945. m. Dr. Eugene Rabinowitch, 935 E. 60th St., Chicago, Ill. 60637

Bulletin of the International Association of Scientific Hydrology. 1956. q. Braamstraat 61, Gantbrugge, Belgium.

California Citrograph. 1915. m. 5380 Poplar Rd., Los Angeles, Calif. 90032.

Computer Journal. 1958. q. British Computer Society, 23 Dorset Sq., London N.W.1, England.

Earth Science Reviews. 1965. q. Elsevier Publishing Co., P.O. Box 211, 335 Jan van Galenstraat, Amsterdam, Netherlands.

Ecological Monographs. 1931. q. Duke University Press, Box 6697, College Station, Durham, N.C. 27708.

Ecology. 1920. 6/yr. Duke University Press, Box 6697, College Station, Durham, N.C. 27708.

Economic Botany. 1947. q. New York Botanical Garden, Bronx, N.Y. 10458.

Economic Geology. 1905. 8/yr. Box 26, Blacksburg, Va. 24060.

Environment and Behavior. 1969. q. Sage Publications, 275 South Beverly Drive, Beverly Hills, Calif. 90212.

Environmental Education, m. DERS, Box 1605, Madison, Wis. 53701.

Environmental Health Letter. 1961. bi-m. 1097 National Press Building, Washington, D.C. 20004.

Environmental Research. 1967. q. Academic Press, 111 Fifth Ave., New York, N.Y. 10003.

Environmental Science & Technology. 1967. m. American Chemical Society, 1155 16th St., N.W., Washington, D.C. 20036.

Environmental Technology and Economics. 1966. bi-w. Technomic Publishing Co., Stamford, Conn.

Forest Science. 1955. q. Society of American Foresters, 1010 16th St., N.W., Washington, D.C. 20036.

Geodesy and Aerophotography. 1962. bi-m. American Geophysical Union, 1145 19th St., N.W., Washington, D.C. 20036.

Geological Magazine. 1864. bi-m. Stephen Austin & Sons, Ltd., Caxton Hill, Ware Rd., Hertford, Hertfordshire, England.

GEOLOGICAL SOCIETY OF AMERICA, *Bulletin*. 1888. m. 231 E. 46th St., New York, N.Y. 10017.

GEOLOGICAL SOCIETY OF LONDON, *Quarterly Journal*. 1845. q. H. K. Lewis & Co., Ltd., 136 Gower St., London W.C.1, England.

GeoScience News. 1967. bi-m. P.O. Box 4428, Pasadena, Calif. 91106.

GeoTimes. 1956. 8/yr. American Geological Institute, 1444 N St., N.W., Washington, D.C. 10005.

Gerlands Breiträge zur Geophysik. 1887. 6/yr. Akademische Verlagsgesellschaft Geest & Portig, K-G, Sterwartenstr. 8, Leipzig C.1, Germany.

Impact of Science on Society. q. United Nations Educational, Scientific and Cultural Organization, Place de Fontenoy, Paris (7^e), France.

International Geology Review. 1959. m. American Geological Institute, 1444 N St., N.W., Washington, D.C., 20005.

International Journal of Bioclimatology and Biometeorology. 1957. 3/yr. Hofbrouckerlaah 54, Oegstgeest, Leiden, Netherlands.

Journal of Applied Ecology. 1964. s-a. (British Ecological Society.) Blackwell Scientific Publications Ltd., 5 Alfred St., Oxford, England.

Journal of Applied Meteorology. 1962. q. American Meteorological Society, 45 Beacon St., Boston, Mass. 02108.

Journal of Atmospheric and Terrestrial Physics. 1950. m. Pergamon Press Ltd., 122 E. 55th St., New York, N.Y. 10022.

Journal of Ecology. 1913. 3/yr. (British Ecological Society.) Blackwell Scientific Publications Ltd., 5 Alfred St., Oxford, England.

Journal of Forestry. 1917. m. Society of American Foresters, 1010 16th St., N.W., Washington, D.C. 20036.

Journal of Geology. 1893. bi-m. University of Chicago Press, 5750 Ellis Ave., Chicago, Ill. 60637.

Journal of Geophysical Research. 1896. m. American Geophysical Union, 1145 19th St., N.W., Washington, D.C. 20036.

Journal of Glaciology. 1949. 3/yr. Scott Polar Research Institute, Lensfield Road, Cambridge, England.

Journal of Hydrology. 1963. 4/yr. North Holland Publishing Co., 68–70 Nieuwezijds Voorburgwal, Box 103, Amsterdam, Netherlands.

Journal of Marine Research. 1937. 3/yr. Box 2025, Yale University, New Haven, Conn. 06520.

Journal of Soil and Water Conservation. 1946. bi-m. Soil Conservation Society of America, 7515 Northeast Ankeny Rd., Ankeny, Iowa 50021.

Journal of the Atmospheric Sciences. 1944. bi-m. American Meteorological Society, 45 Beacon St., Boston, Mass. 02108.

Limnology and Oceanography. 1956. q. University of Michigan, Ann Arbor, Mich. 49060.

Meteorological Magazine. 1866. m. H.M.S.O. Atlantic House, Holborn Viaduct, London E.C.1, England.

Meteorological Society of Japan, *Journal.* 1882. bi-m. Charles E. Tuttle Co., Tokyo, Japan.

Monthly Weather Review. 1872. m. Superintendent of Documents, Washington, D.C. 20402.

Natural History. 1900. 10/yr. American Museum of Natural History, Central Park West at 79th St., New York, N.Y. 10024.

Natural Resources Journal. 1961. q. University of New Mexico, School of Law, 1915 Roma N.E., Albuquerque, N. Mex. 87106.

Nature. 1869. w. Macmillan (Journals), Ltd., Little Essex St., London W.C.2, England.

Nature and Resources. 1958. q. (Formerly *Arid Zone*) United Nations Educational, Scientific and Cultural Organization, Place de Fontenoy, Paris (7ᵉ), France.

Pacific Science. 1947. q. University of Hawaii Press, Honolulu, Hawaii 96822.

Palaeogeography, Palaeoclimatology, Palaeoecology. 1965. q. Elsevier Publishing Co., Box 211, 335 Jan van Galenstraat, Amsterdam, Netherlands.

Photogrammetric Engineering. 1934. 6/yr. American Society of Photogrammetry, 105 N. Virginia Ave., Falls Church, Va. 22046.

Polar Record. 1955. 3/yr. Scott Polar Research Institute, Lensfield Road, Cambridge, England.

Reclamation Era. 1908. q. Bureau of Reclamation, Washington, D.C. 20402.

Remote Sensing of Environment. 1969. q. American Elsevier Publishing Co., New York, N.Y.

ROYAL METEOROLOGICAL SOCIETY, *Quarterly Journal.* 1871. q. 49 Cromwell Road, London S.W.7, England.

Science. 1880. w. American Association for the Advancement of Science, 1515 Massachusetts Ave., N.W., Washington, D.C. 20005.

Scientific American. 1845. m. 415 Madison Ave., New York, N.Y. 10017.

Sedimentology. 1961. 8/yr. Elsevier Publishing Co., Box 211, 355 Jan van Galanstraat, Amsterdam, Netherlands.

SEISMOLOGICAL SOCIETY OF AMERICA, *Bulletin.* 1911. bi-m. Waverly Press, Inc., Baltimore, Md.

Sierra Club Bulletin. 1925. 10/yr. 1050 Mills Tower, San Francisco, Calif.

Soil Conservation. 1935. m. (Soil Conservation Service.) Superintendent of Documents, Washington, D.C. 20402.

Soil Science. 1916. m. Williams & Wilkins Co., 428 E. Preston St., Baltimore, Md. 21202.

Solar Energy. 1957. q. Arizona State University, Tempe, Ariz.

Southwestern Naturalist. 1953. 4/yr. The Southwestern Association of Naturalists, Department of Biology, Texas Technological College, Lubbock, Tex. 79406.

Soviet Hydrology. 1963. 6/yr. American Geophysical Union, 1145 19th St., N.W., Washington, D.C. 20005.

Surveying and Mapping. 1941. q. Congress on Surveying and Mapping, Box 470, Benjamin Franklin Station, Washington, D.C. 20044.

Tellus. Svenska Geofysiska Foreningen, Tulegatan 41, Stockholm VA, Sweden.

Tree-Ring Bulletin. 1935. q. Laboratory of Tree-Ring Research, University of Arizona, Tucson, Ariz. 85721.

Water Resources Bulletin. 1965. q. American Water Resources Association, 103 North Race St., Urbana, Ill. 61801.

Water Resources Research. 1965. Bi-m. American Geophysical Union, Suite 435, 2100 Pennsylvania Ave., N.W., Washington, D.C. 20037.

Weather. 1946. m. Royal Meteorological Society, 49 Cromwell Rd., London S.W.7, England.

Weatherwise. 1948. bi-m. American Meteorological Society, 45 Beacon St., Boston, Mass. 02108.

WMO Bulletin. 1952. q. World Meteorological Organization, 41 av. Giuseppe Motta, Geneva, Switzerland.

Zeitschrift für Geomorphologie. New Series 1957. 4/yr. Gebrüder Borntraeger, An der Rehwiese 14, Berlin-Nikolassee, Germany.

C. Social Sciences

African Affairs. 1901. q. Royal African Society, 18 Northumberland Ave., London W.C.2, England.

AFRICAN STUDIES ASSOCIATION OF THE UNITED KINGDOM, *Bulletin.* 1964. 3/yr. % Centre of West African Studies.

Agricultural History. 1927. q. University of California Press, Berkeley, Calif. 94720.

AMERICAN ACADEMY OF POLITICAL AND SOCIAL SCIENCE, *Annals.* 1890. bi-m. 3937 Chestnut Street, Philadelphia, Pa. 19104.

American Anthropologist. 1888. 6/yr. 3700 Massachusetts Ave., N.W., Washington, D.C. 20016.

American Antiquity. 1935. q. Society for American Archaeology, 3700 Massachusetts Ave., Washington, D.C. 20016.

American City Magazine. 1909. m. Buttenheim Publishing Corp., 470 Park Ave. So., New York, 10017.

American Highways. 1922. q. National Press Bldg., Washington, D.C. 20004.

American Historical Review. 1895. q. Macmillan Co., 866 Third Ave., New York, N.Y. 10022.

AMERICAN INDUSTRIAL DEVELOPMENT COUNCIL, *Journal.* 1966. q. AIDC, 230 Boylston St., Boston, Mass. 02116.

AMERICAN INSTITUTE OF ARCHITECTS, *Journal.* 1944. m. The Octagon, 1735 New York Ave., N.W., Washington, D.C. 20006.

AMERICAN INSTITUTE OF PLANNERS, *Journal.* 1925. 6/yr. 917 15th St., N.W., Washington, D.C. 20055.

American West. 1964. q., 577 College Ave., Palo Alto, Calif. 94306.

The Annals of Regional Science. 1967. Bellingham, Wash. (Western Regional Science Association.)

Antiquity. 1927. q. W. Heffer & Sons Ltd., Cambridge, England.

Arab Journal. q. Organization of Arab Students in the U.S.A. and Canada, 2929 Broadway, New York, N.Y. 10026.

Archaeology. 1948. q. Archaeological Institute of America, 100 Washington Sq. E., New York, N.Y. 10003.

Arctic Anthropology. 1962. irreg. University of Wisconsin Press, Madison, Wis. 53706.

Asian Review of Art and Letters. 1964. 3/yr. 2 Temple Chambers, Temple Ave., London E.C.4, England.

Asian Survey. 1961. m. University of California, Institute of International Studies, 2538 Channing Way, Berkeley, Calif. 94720.

California Farmer. 1854. fortn. 83 Stevenson St., San Francisco, Calif. 94105.

The Canadian Cartographer. 2/yr. Department of Geography. York University, 4700 Keele So., Downsview, Toronto, Canada.

Canadian Historical Review. 1920. q. University of Toronto Press, Toronto, Canada.

Cartography. 1/yr. Australian Institute of Cartographers, Box 1020 H, G.P.O., Elizabeth St., Melbourne, Victoria, Australia.

The Center Magazine. 1967. bi-m. John Cogley, Center for the Study of Democratic Institutions, Box 4068, Santa Barbara, Calif.

Central Asian Review. 1953. q. Central Asian Research Centre, 66a Kings Rd., London S.W.3, England.

Cry California. 1965. q. California Tomorrow, San Francisco, Calif.

Current Anthropology. 1960. 5/yr. University of Chicago, 1126 E. 59th Street, Chicago, Ill. 60637.

Current History. 1914. m. 1822 Ludlow St., Philadelphia, Pa. 19103.

Daedalus. 1955. q. American Academy of Arts and Sciences, 280 Newton St., Brookline Sta., Boston, Mass. 02146.

Demography. 1964. 2/yr. The Population Association of America, 1126 East 59th St., University of Chicago, Chicago, Ill. 60637.

East European Quarterly. 1967. q. University of Colorado, Regent Hall, Boulder, Colo. 80302.

Economic Development and Cultural Change. 1952. q. University of Chicago Press, 5750 Ellis Ave., Chicago, Ill. 60637.

Economic History Review. 1927. 3/yr. Broadwater Press Ltd., Welwyn Garden City, Herts, England.

Ekistica. 1955. m. Athens Technological Institute, 24 Strat, Syndesinov, Athens, 136, Greece.

Ethnohistory. American Society for Ethnohistory, 4242 Ridge Lea Rd., Amhurst, New York 14226.

Ethnology. 1962. q. University of Pittsburgh, Department of Anthropology, 4200 Fifth Ave., Pittsburgh, Pa. 15213.

FAO Review. 1968. 6/yr. United Nations, Rome, Italy.

Farm Quarterly. 1946. 6/yr. F & W Publishing Corp., 22 E. 12th St., Cincinnati, Ohio 45210.

Foreign Agriculture. (Foreign Agricultural Service, U.S. Department of Agriculture.) 1937. m. Superintendent of Documents, Washington, D.C. 20402.

Growth and Change. 1970. College of Business and Economics, University of Kentucky, Lexington, Ky. 40506.

Hispanic American Historical Review. 1918. q. Duke University Press, Box 6697, College Station, Durham, N.C. 27708.

Historia Mexicana. 1951. q. Colegio de Mexico, Guanajuato 125, Mexico 7, D.F.

HISTORICAL SOCIETY OF SOUTHERN CALIFORNIA, *Publications.* 1884. q. 1909 S. Western Ave., Los Angeles, Calif. 90018.

Historical Studies. (Formerly *Historical Studies: Australia and New Zealand.*) 1940. s-a. University of Melbourne, Department of History, Parkville, N.2., Victoria, Australia.

Human Organization. 1941. q. Society for Applied Anthropology, University of Kentucky, Lexington, Ky. 40506.

Indo-Asian Culture. 1952. q. Indian Council for Cultural Relations, Azad Bhavan, Indraprastha Estate, New Delhi 1, India.

INSTITUTE FOR THE STUDY OF THE USSR, *Bulletin.* 1954. m. Instituto de Estudios Politicos, Plaza de la Marina Espanola, 8 Madrid, Spain.

International Economic Review. 1960. 3/yr. Kansai Economic Federation, Shin-Osaka Bldg., Dojima-Hamadori, Keta-ku, Osaka, Japan.

International Migration Review. (Formerly *International Migration Digest.*) 1964. 3/yr. Center for Migration Studies, 209 Flagg Place, Staten Island, N.Y. 10304.

International Organization. 1947. q. World Peace Foundation, 40 Mt. Vernon St., Boston, Mass. 02108.

International Review of Social History. 1956. 3/yr. Royal Van Gorcum Ltd., Assen, Netherlands.

International Social Science Journal/ Revue Internationale des Sciences Sociales. 1949. q. United Nations Educational, Scientific and Cultural Organization, Place de Fontenoy, Paris (7e), France.

ISIS. 1913. q. Johns Hopkins University, Baltimore, Md. 21218.

Journal of African History. 1960. 3/yr. Cambridge University Press, 200 Euston Rd., London N.W.1, England. (32 E. 57th St., New York, N.Y. 10022.)

Journal of Asian and African Studies. 1966. q. E. J. Brill, Leiden, Netherlands.

Journal of Asian Studies. 1941. 5/yr. Duke University, Department of History, Durham, N.C. 27708.

Journal of Development Studies. 1964. q. Frank Cass & Co., Ltd., 67 Great Russell St., London W.C.1, England.

Journal of Economic History. 1941. q. Economic History Association, 100 Trinity Place, New York, N.Y. 10006.

Journal of Farm Economics. 1919. 5/yr. American Farm Economic Association, Cornell University, Ithaca, N.Y. 14850.

Journal of Inter-American Studies. 1959. q. University of Miami Press, Box 8134, University of Miami, Coral Gables, Fla. 33124.

Journal of Latin American Studies. 1969. Cambridge University Press, Bentley House, 200 Euston Road, London N.W.1, England.

Journal of Leisure Research. 1969. National Recreation and Park Association, 1700 Pennsylvania Ave., N.W., Washington, D.C. 20006.

Journal of Modern African Studies. 1963. q. Cambridge University Press, 200 Euston Rd., London N.W.1, England.

Journal of Near Eastern Studies. 1884. q. University of Chicago Press, 5750 Ellis Ave., Chicago, Ill. 60637.

Journal of Negro History. 1916. q. Association for the Study of Negro Life and History, Inc., 1538 Ninth St., N.W., Washington, D.C. 20001.

Journal of Regional Science. 1958. Regional Science Research Institute, Box 8776, Philadelphia, 19101.

Journal of Social History. 1967. q. University of California Press, 2223 Fulton St., Berkeley, Calif. 94720.

Journal of Social Issues. 1945. q. Society for the Psychological Study of Social Issues, Box 1248, Ann Arbor, Mich. 48106.

Journal of Southeast Asian History. 1960. 2/yr. Department of History, University of Singapore.

Journal of Southern History. 1935. q. Southern Historical Association, Tulane University, New Orleans, La. 70118.

Journal of the Economic & Social History of the Orent Jesho. 1957. 3/yr. E. J. Brill N.V., Leiden, Netherlands.

Journal of the History of Ideas. 1940. q. Box 285, City College, New York, N.Y. 10031.

Journal of the Science of Food and Agriculture. 1950. m. Society of Chemical Industry, 14 Belgrave Sq., London S.W.1, England.

Journal of the West. 1962. q. 1915 S. Western Ave., Los Angeles, Calif. 90018.

Kansas Historical Quarterly. 1931. q. Kansas State Historical Society, Memorial Bldg., 120 W. Tenth, Topeka, Kans. 66612.

Land Economics. 1925. q. University of Wisconsin, Madison, Wis. 53706.

Landscape. 1951. 3/yr. Box 7177, Landscape Substation, Berkeley, Calif. 94717.

Landscape Architecture. 1910. q. American Society of Landscape Architects, Schuster Bldg., 1500 Bardstown Rd., Louisville, Ky. 40205.

Latin American Research Review. 1965. 3/yr. University of Texas, Latin American Research Review Board, Box L, Austin, Tex. 78712.

L'Homme. 1961. 4/yr. Mouton Cie., 45 rue de Lille, Paris (7e), France.

LONDON UNIVERSITY, INSTITUTE of HISTORICAL RESEARCH, *Bulletin.* 1923. 2/yr. Senate House, London W.C.1, England.

Man. 1966. q. Royal Anthropological Institute, 21 Bedford Sq., London W.C.1, England.

Middle East Journal. 1946. q. Middle East Institute, 1761 N St., N.W., Washington, D.C. 20036.

Middle Eastern Affairs. 1950. m. Council for Middle Eastern Affairs, Inc., 2061 Belmont Ave., Elmont, N.Y.

Middle Eastern Studies. 1964. q. Frank Cass & Co., Ltd., 67 Great Russell St., London W.C.1, England.

Milbank Memorial Fund Quarterly. 1923. q. 40 Wall St., New York, N.Y. 10005.

Monthly Labor Review. 1915. m. (U.S. Bureau of Labor Statistics.) Superintendent of Documents, Washington, D.C. 20402.

Names. 1953. q. (American Name Society.) State University College, Potsdam, N.Y. 13676.

Near East Report. 1965. fortn. Near East Report, Inc., 1341 G St., N.W., Washington, D.C. 20005.

New Mexico Historical Review. 1926. q. University of New Mexico Press, Albuquerque, N. Mex. 87106.

Oceania. 1930. q. University of Sydney, New South Wales, Australia.

Oregon Historical Quarterly. 1900. q. Oregon Historical Society, 1230 S.W. Park Ave., Portland, Ore. 97205.

Oriental Economist. 1934. m. Nehombashi, Tokyo, Japan.

Pacific Affairs. 1928. q. University of British Columbia, Vancouver 8, Canada.

Pacific Discovery. 1948. bi-m. California Academy of Sciences, Golden Gate Park, San Francisco 18, Calif.

Pacific Historian. 1957. q. University of the Pacific, Stockton, Calif. 95024.

Pacific Historical Review. 1932. q. University of California Press, Berkeley, Calif. 94720.

Pacific Northwest Quarterly. 1906. q. University of Washington, Parrington Hall, Seattle, Wash. 98105.

Pacific Sociological Review. 1958. s-a. John M. Foskett. University of Oregon, Department of Sociology, Eugene, Ore. 97403.

Political Science Quarterly. 1886. q. Dr. Sigmund Diamond. Columbia University, Academy of Political Science, Fayerweather Hall, New York, N.Y. 10027.

Population. 1946. bi-m. Institut National d'Etudes Demographiques (INED), 23 av. Franklin Roosevelt, Paris (8^e), France.

Population Bulletin. 1945. 5/yr. Population Reference Bureau, Inc., 1755 Massachusetts Ave., N.W., Washington D.C. 20036.

Population Studies. 1947. 3/yr. (1 vol. per yr.) D. V. Glass & E. Grebenik. London School of Economics, Population Investigation Committee, Houghton St., London W.C.2, England.

Polynesian Society, *Journal.* 1892. q. Ralph Bulmer, Polynesian Society, Box 5195, Wellington, C.1, New Zealand.

Race. 1959. s-a. Oxford University Press, Ely House, 37 Dover St., London W.1, England.

Review of Economic Studies. 1933. q. Oliver and Boyd Ltd., Tweeddale Court, 14 High St., Edinburgh 1, Scotland.

Revista Iberoamericana. 1938. s-a. Editorial Cultura (Mexico), Republica de Guatemala 96, Mexico 1, D.F.

Revista Internacional de Sociologio. 1943. q. Instituto "Balmes" de Sociologia, Universidad de Madrid, Madrid, Spain.

Revue Historique. 1876. q. Presses Universitaires de France, Département des Periodiques, 12 Rue Jean-de-Beauvais, 75 Paris (5e), France.

Royal Central Asian Society, *Journal.* 1903. 3/yr. Royal Central Asian Society, 42 Devonshire St., London W.1, England.

Rural Sociology. 1936. q. Rural Sociology Society, University of Wisconsin, Madison, Wis.

Scandinavian Economic History Review. 1953. 2/yr. Institute of Economic History, Bispetorvet 3, DK-1167, Copenhagen K, Denmark.

Science and Culture. 1964. m. Indian Science News Association, 92 Acharya Prafullachandra Rd., Calcutta 9, India.

Sinologica; Zeitschrift für chinesische Kultur und Wissenschaft. 1947. irreg. Verlag für Recht und Gesellschaft AG., Bundesstr. 15, Basel, Switzerland.

Slavic Review. 1941. q. Columbia University, 622 W. 113th St., New York, N.Y. 10025.

Smithsonian Journal of History. 1966. q. Smithsonian Institution, Washington, D.C. 20560.

Social and Economic Studies. 1953. m. University of the West Indies, Institute of Social and Economic Research, Mona, Jamaica.

Social Science Quarterly. 1920. q. Charles M. Bonjean, University of Texas at Austin, Austin, Tex. 78712. Published originally as *Southwest Social Science Quarterly.*

Sociological Quarterly. 1960. 4/yr. Paul J. Campisi. Midwest Sociological Society, Department of Sociology, Southern Illinois University, Edwardsville, Ill. 62025.

Sociological Review. 1908. 3/yr. University of Keele, Keele, Staffordshire, England.

South African Journal of Economics. 1933. q. Economic Society of South Africa, Box 5316, Johannesburg, South Africa.

Southwestern Historical Quarterly. 1897. q. Texas State Historical Association, Box 8059, University Station, Austin, Tex. 78712.

Southwestern Journal of Anthropology. 1945. q. University of New Mexico, Albuquerque, N. Mex. 87106.

Soviet Anthropology and Archaeology. 1962. q. International Arts & Sciences Press, 108 Grand St., White Plains, N.Y. 10601.

Studies on the Soviet Union. 1957. q. Institute for the Study of the USSR, Mannhardtstr. 6, 8 Munich 22, W. Germany.

Technology and Culture. 1960. q. Society for the History of Technology, University of Chicago Press, 5750 Ellis Ave., Chicago, Ill. 60637.

Town and Country Planning. 1932. m. Town and Country Planning Association, The Planning Centre, 28 King St., London W.C.2, England.

TOWN PLANNING INSTITUTE, *Journal.* 1914. 10/yr. Town Planning Institute, 26 Portland Place, London W.1, England.

Town Planning Review. 1910. q. Liverpool University Press, 123 Grove Street, Liverpool 7, England.

Traffic Quarterly. 1947. q. Eno Foundation for Highway Traffic Control, Inc., Saugatuck, Conn. 06880.

Transportation Journal. 1961. q. American Society of Traffic and Transportation, Inc., 22 W. Madison St., Chicago, Ill. 60602.

Transportation Research. 1967. q. Pergamon Press Inc., Maxwell House, Fairview Park, Elmsford, N.Y. 10523.

Transportation Science. 1967. q. Operations Research Society of America, 428 East Preston St., Baltimore, Md. 21202.

Tropical Agriculture. 1924. q. Butterworths Scientific Publications, 88 Kingsway, London W.C.2, England.

United Asia. 1948. bi-m. 12 Rampart Row, Bombay 1, India.

Urban Affairs Quarterly. 1965. q. Sage Publications, Inc., 275 S. Beverly Drive, Beverly Hills, Calif. 90212.

Urban Land. 1941. 11/yr. Urban Land Institute, 1200 18th St., N.W., Washington, D.C. 20036.

Urban Research News. 1966. fortn. Sage Publications, Inc., 275 S. Beverly Drive, Beverly Hills, Calif. 90212.

Urban Studies. 1964. 3/yr. Oliver & Boyd Ltd., Tweeddale Ct., 14 High Street, Edinburgh 1, Scotland.

Urban West. bi-m. 593 Market St., San Francisco, Calif. 94105.

U.S. FOREIGN AGRICULTURAL SERVICE, *World Agricultural Production & Trade.* 1965. m. Superintendent of Documents, Government Printing Office, Washington, D.C. 20402.

Utah Historical Quarterly. 1928. q. Utah State Historical Society, 603 E. South Temple, Salt Lake City, Utah 84102.

West Africa. 1917. w. Overseas Newspapers (agencies) Ltd., 9 New Fetter Lane, London E.C.4, England.

Western City. 1925. m. League of California Cities, 702 Statler Center, Los Angeles, Calif. 90017.

Western Economic Journal. 1962. 4/yr. Harold Somers and Alice Vandermeulan, University of California, Department of Economics, Los Angeles, Calif. 90024.

Western Folklore. 1942. q. (California Folklore Society.) Albert B. Friedman, University of California Press, Berkeley, Calif. 94720.

W.H.O. Chronicle. 1947. m. World Health Organization, Sales Section, Geneva, Switzerland.

William and Mary Quarterly. 1892. q. Institute of Early American History and Culture, Box 220, Williamsburg, Va. 23185.

Index of Geographical Reference Works as Cited in Chapters 5, 6, and 7

Subject Index